Advanced Information and Knowledge Processing

Information systems and intelligent knowledge processing are playing an increasing role in business, science and technology. Recently, advanced information systems have evolved to facilitate the co-evolution of human and information networks within communities. These advanced information systems use various paradigms including artificial intelligence, knowledge management, and neural science as well as conventional information processing paradigms.

The aim of this series is to publish books on new designs and applications of advanced information and knowledge processing paradigms in areas including but not limited to aviation, business, security, education, engineering, health, management, and science.

Books in the series should have a strong focus on information processing - preferably combined with, or extended by, new results from adjacent sciences. Proposals for research monographs, reference books, coherently integrated multi-author edited books, and handbooks will be considered for the series and each proposal will be reviewed by the Series Editors, with additional reviews from the editorial board and independent reviewers where appropriate. Titles published within the Advanced Information and Knowledge Processing Series are included in Thomson Reuters' Book Citation Index and Scopus.

More information about this series at http://www.springer.com/series/4738

Leslie F. Sikos • Oshani W. Seneviratne
Deborah L. McGuinness

Editors

Provenance in Data Science

From Data Models to Context-Aware
Knowledge Graphs

 Springer

Editors
Leslie F. Sikos ⓘ
Edith Cowan University
Perth, WA, Australia

Oshani W. Seneviratne ⓘ
Health Data Research
Rensselaer Polytechnic Institute
Troy, NY, USA

Deborah L. McGuinness ⓘ
Rensselaer Polytechnic Institute
Troy, NY, USA

ISSN 1610-3947   ISSN 2197-8441   (electronic)
Advanced Information and Knowledge Processing
ISBN 978-3-030-67683-4   ISBN 978-3-030-67681-0   (eBook)
https://doi.org/10.1007/978-3-030-67681-0

This Springer imprint is published by the registered company Springer Nature Switzerland AG
The registered company address is: Gewerbestrasse 11, 6330 Cham, Switzerland

Preface

Since their popularization by Google in 2012, knowledge graphs became the new powerhouse for representing interconnected information of various knowledge domains. What is common in these graphs is that they all use a graph-based data model. The encoding language, style, and infrastructures may vary. Ultimately, knowledge graphs contain information about things and their interrelationships, and the content is represented with something like the Resource Description Framework (RDF). Knowledge graphs broadly cover graphical encodings of semantics and thus include the space of conceptual graphs, labeled property graphs, and hypergraphs. They are often stored in graph databases or triplestores. Knowledge graphs may be seen as marking a new era in data mining, integration, and visualization. They are also powering a range of applications that require distributed processing, complex querying, and automated reasoning. This is witnessed by the growing number of software implementations and feature support—see the now available RDF knowledge graph support in the latest releases of the industry-leading Oracle Database, which was originally designed as a purely relational database, for example; the Oracle Spatial and Graph support for Graph Visualization using Cytoscape; or the Amazon Neptune fully managed, scalable graph database, which supports both the RDF data model and the property graph model of Apache TinkerPop.

The structure of data determines the tasks that software agents can perform on the data. Knowledge graphs are gaining traction in areas including natural language processing and machine learning implementations as those applications can leverage the structured or semi-structured content captured in the knowledge graphs. Graph-based knowledge representations are highly accessible, scalable, and easy to visualize. They support advanced, semantic search, automated recommendation, and various data querying options corresponding to graph traversal algorithms. While there are many research challenges, knowledge graphs already play an increasingly important role in the development of a range of hybrid AI systems that significantly outperform traditional data processing systems, such as via enhanced deep learning based on knowledge-infused learning processes.

The most significant knowledge graph to date, the World Wide Web, has seen its fair share of challenges and opportunities in its three-plus decades of existence.

Users have linked to other documents that resulted in one of the first forms of rudimentary knowledge graphs. Wikipedia started crowd-sourcing knowledge on the Web to create the largest encyclopedia ever to be compiled with many links and connections, resulting in a versatile knowledge graph. The research and development work on Semantic Web has given rise to more structured data formats and even richer data sources on the Web, such as DBPedia and WikiData. Online social networks connected individuals worldwide, giving rise to yet another very large-scale knowledge graph. With the increasing volume of information, questions about where things are coming from became one of the fundamental driving forces behind RDF-based provenance expression, ultimately resulting in a W3C recommendation for provenance.

This book provides a collection of articles on provenance techniques and state-of-the-art metadata-enhanced, provenance-aware, graph-based representations across multiple application areas. Provenance is particularly important in the medical sciences and scientific literature to support claims by capturing information origin/data source, timestamps, agents generating the data, and so on. Provenance is also important in heterogeneous environments, such as communication networks and cyberspace, in which data is typically derived from diverse sources and obtained via data aggregation or data fusion. Therefore, provenance is of utmost importance, making a distinction between reliable and unreliable data, or a harmless online action and a cyberattack. The combined use of graph-based data models and provenance representations may be seen as providing the best of both worlds, as described in this book.

Perth, WA, Australia   Leslie F. Sikos

Troy, NY, USA   Oshani W. Seneviratne

Troy, NY, USA   Deborah L. McGuinness

March 2021

Contents

Contents

About the Editors

Dr. Leslie F. Sikos is a computer scientist specializing in artificial intelligence and data science, with a focus on cybersecurity applications. He holds two Ph.D. degrees and 20+ industry certificates. He is an active member of the research community as an author, editor, reviewer, conference organizer, and speaker, and a member of industry-leading organizations, such as the ACM and the IEEE. He contributed to international standards and developed state-of-the-art knowledge organization systems. As part of a major cybersecurity research project, he co-authored a formalism for capturing provenance-aware network knowledge with RDF quadruples. He has published more than 20 books, including textbooks, monographs, and edited volumes.

Dr. Oshani W. Seneviratne is the Director of Health Data Research at the Institute for Data Exploration and Applications at the Rensselaer Polytechnic Institute (Rensselaer IDEA). She obtained her Ph.D. in Computer Science from the Massachusetts Institute of Technology in 2014 under the supervision of Sir Tim Berners-Lee, the inventor of the World Wide Web. During her Ph.D., Oshani researched Accountable Systems for the Web. She invented a novel web protocol called HTTPA (HyperText Transfer Protocol with Accountability), and a novel provenance tracking mechanism called the Provenance Tracking Network. This work was demonstrated to be effective in several domains including electronic health care records transfer, and intellectual property protection in Web-based decentralized systems. At the Rensselaer IDEA, Oshani leads the Smart Contracts Augmented with Analytics Learning and Semantics (SCALeS) project. The goal of this project is to predict, detect, and fix initially unforeseen situations in smart contracts utilizing novel combinations of machine learning, program analysis, and semantic technologies. Oshani is also involved in the Health Empowerment by Analytics, Learning, and Semantics (HEALS) Project. In HEALS, she oversees the research operations targeted at the characterization and analysis of computational medical guidelines for chronic diseases such as diabetes, and the modeling of guideline provenance. Before Rensselaer, Oshani worked at Oracle specializing in distributed systems, provenance and healthcare-related research. She is the co-inventor of two enterprise provenance patents.

Prof. Deborah L. McGuinness is the Tetherless World Senior Constellation Chair and Professor of Computer, Cognitive, and Web Sciences at RPI. She is also the founding director of the Web Science Research Center and the CEO of McGuinness Associates Consulting. Deborah has been recognized with awards as a fellow of the American Association for the Advancement of Science (AAAS) for contributions to the Semantic Web, knowledge representation, and reasoning environments and as the recipient of the Robert Engelmore Award from the Association for the Advancement of Artificial Intelligence (AAAI) for leadership in Semantic Web research and in bridging Artificial Intelligence (AI) and eScience, significant contributions to deployed AI applications, and extensive service to the AI community. Deborah leads a number of large diverse data-intensive resource efforts, and her team is creating next-generation ontology-enabled research infrastructure for work in large interdisciplinary settings. Prior to joining RPI, Deborah was the acting director of the Knowledge Systems, Artificial Intelligence Laboratory, and Senior Research Scientist in the Computer Science Department of Stanford University, and previous to that, she was at AT&T Bell Laboratories. Deborah consults with numerous large corporations as well as emerging startup companies wishing to plan, develop, deploy, and maintain Semantic Web and/or AI applications. Some areas of recent work include: data science, next-generation

health advisors, ontology design and evolution environments, semantically enabled virtual observatories,
semantic integration of scientific data, context-aware
mobile applications, search, eCommerce, configuration,
and supply chain management. Deborah holds a bache-
lor's degree in Math and Computer Science from Duke
University, a master's degree in Computer Science from
the University of California at Berkeley, and a Ph.D. in
Computer Science from Rutgers University.

Chapter 1
The Evolution of Context-Aware RDF Knowledge Graphs

Leslie F. Sikos ⓘ

1.1 Introduction to RDF Data Provenance

The power of the *Resource Description Framework* (RDF)[1] lies in its simplicity in a way that it allows making machine-interpretable statements in the form of subject-predicate-object triples, formally $(s, p, o) \in (\mathbb{I} \cup \mathbb{B}) \times \mathbb{I} \times (\mathbb{I} \cup \mathbb{L} \cup \mathbb{B})$, where $\mathbb{I}$ is the set of Internationalized Resource Identifiers (IRIs), that is, strings of Unicode characters of the form `scheme:[//[user:password@]` `host[:port]][/]path[?query][#fragment]`, or any valid subset of these (in particular, URIs such as URLs); $\mathbb{L}$ is the set of RDF literals that can be 1) self-denoting plain literals $\mathbb{L}_P$ of the form `"<string>"(@<lang>)?`, where `<string>` is a string and `<lang>` is an optional language tag, or 2) typed literals $\mathbb{L}_T$ of the form `"<string>"|datatype>`, where `<datatype>` is an IRI denoting a datatype according to a schema and `<string>` is an element of the lexical space corresponding to the datatype; and $\mathbb{B}$ is the set of blank nodes (bnodes), which are unique resources that are neither IRIs nor RDF literals (Sikos 2017b). Sets of such triples form RDF graphs in which the set of nodes is the set of subjects and objects of RDF triples, and the edges are the predicates of RDF triples.

RDF triples (also known as RDF statements) can be used to define terminological, assertional, and relational axioms in knowledge organizational systems such as knowledge bases and ontologies (Sikos 2017c); enable complex querying, in particular with the purpose-design SPARQL query language (Sikos 2017a); and facilitate automated reasoning to make implicit dataset statements explicit.

---

[1] https://www.w3.org/RDF/

L. F. Sikos (✉)
Edith Cowan University, Perth, WA, Australia
e-mail: l.sikos@ecu.edu.au

© Springer Nature Switzerland AG 2021
L. F. Sikos et al. (eds.), *Provenance in Data Science*, Advanced Information and Knowledge Processing, https://doi.org/10.1007/978-3-030-67681-0_1

Arguably though, many applications require more than just triples, such as metadata, provenance data, or general context information, to provide data origin, authorship, recency, or certainty. Among these, provenance is particularly important as it reveals the origin of derived information, thereby enabling the assessment of information quality, judging trustworthiness and accountability, and the analysis of information on the spatiotemporal plane (which can also be complemented by purpose-designed knowledge representation mechanisms (Sikos 2018b,a; Sikos et al. 2018)).

Because of the nature and range of serialization formats of the data created, transferred, shared, and constantly updated on today's interconnected networks, RDF can be a straightforward choice for all the applications in which bridging data heterogeneity gaps is highly desired (Sikos et al. 2018). However, by design, there is no built-in mechanism in RDF to capture data provenance. Only two techniques have been proposed by the W3C, the standardization body behind RDF: *RDF reification*,[2] which makes RDF statements about other RDF statements by instantiating the `rdf:Statement` class and using the `rdf:subject`, `rdf:predicate`, and `rdf:object` properties of the standard RDF vocabulary[3] for identifying the elements of the triple; and *n-ary relations*,[4] which is capable of describing the relation between individuals or between an individual and a data value via a purposefully created predicate. Both techniques have several well-known issues. Among these, having a large increase in the number of triples (known as *triple bloat*) that leads to scalability issues, reliance on blank nodes, not defined formal semantics, poor support for querying and reasoning over provenance, and application dependence are the most important (Sikos and Philp 2020).

To be able to utilize RDF and complement standard triples with useful information, several alternatives have been introduced over the years, which address the limitations of RDF from various viewpoints. These are detailed in the following sections.

1.2 Extensions of the Standard RDF Data Model

N3Logic, proposed by Berners-Lee et al. (2008), can capture RDF data provenance by allowing statements to be made about, and to query, other statements using quoted formulae. To achieve this, N3Logic extends RDF in two ways: by providing a syntax extension and by defining a vocabulary of purpose-designed predicates.

RDF+ is a formalized, generic approach for managing various dimensions of formally represented knowledge, such as data source, authorship, certainty, and how up-to-date statements are (Dividino et al. 2009). This approach reuses the modeling capabilities of standard RDF, but extends SPARQL query processing to support the segregation of data and metadata querying.

---

[2]https://www.w3.org/TR/rdf-primer/#reification

[3]https://www.w3.org/1999/02/22-rdf-syntax-ns.ttl

[4]https://www.w3.org/TR/swbp-n-aryRelations/

SPOTL(X) is a knowledge representation model that extends the subject-predicate-object of RDF triples with time and location, thereby providing a spatiotemporal enhancement for factual data expressed in RDF (forming quintuples), and an optional context component to include keywords or keyphrases (leading to sextuples) (Hoffart et al. 2013).

SPOTLX tuples support four types of fact representations for browsing spatially, temporally, or contextually enhanced facts from a compliant knowledge base, such as YAGO2:

- facts of the form *(id, s, p, o)* identified by identifier *id* and described using regular subject-predicate-object triples;
- facts of the form *(id, t_b, t_e)* identified by identifier *id* and associated with a time interval *[t_b, t_e]*;
- facts of the form *(id, lat, lon)* identified by identifier *id* and associated with geographical coordinates (latitude-longitude pairs);
- facts of the form *(id, c)* identified by identifier *id* and associated with context *c*.

To facilitate spatiotemporally and contextually enhanced querying using SPOTLX tuples, various predicates have been introduced for time (overlaps, during, before, after), space (westOf, northOf, eastOf, southOf, nearby), and context (matches).

Hartig and Thompson (2014, 2019) proposed an alternate approach to RDF reification called *RDF**, which embeds triples into (metadata) triples as the subject or object. This requires several extensions to RDF:

- The formal definition of RDF* triples (and RDF* graphs).
- The formal definition of RDF* semantics.
- The extension of at least one of the RDF serialization formats to support delimiters, such as the suggested << and >>, to enclose triples in other triples. For this purpose, the authors proposed *Turtle** as an extension to the standard Turtle syntax.
- An extension to SPARQL syntax, semantics, and grammar to be able to query RDF*-aware graphs. Such an extension is *SPARQL**.

Based on RDF*, statement-level annotations can be made in *RDF streams*,[5] which is called *RSP-QL** (Keskisärkkä et al. 2019).

1.3 Extensions of RDFS and OWL

The extension of RDF's vocabulary for defining terms of a knowledge domain with relationships between them, and datatypes that restrict the permissible value ranges

---

[5]Sequences of RDF graphs, including associated time-related metadata, such as production time, receiving time, start time, end time.

of their properties, is called *RDF Schema (RDFS)*.[6] *Annotated RDF Schema* is a generalized RDF annotation framework that extends the standard RDFS semantics and provides an extension to SPARQL called AnQL (Zimmermann et al. 2011). The annotations can be taken from a specific domain to suit the application, such as to support metadata for temporal, provenance, or fuzzy annotations.

General RDF (g-RDF) triples are positive and strongly negated RDF triples, with a subject that can be a literal and a property that can be a variable (Analyti et al. 2014). These, combined with a g-RDF program that contains derivation rules (whose body is a conjunction of g-RDF triples and scoped weakly negated g-RDF triples, and whose head is a g-RDF triple), form g-RDF ontologies. G-RDF representations support RDF why-provenance to keep track of the origins of derived RDF triples, because derived information based on distributed g-RDF ontologies is associated with the sets of names of g-RDF ontologies that contributed to the derivation of information. This approach utilizes ontologies with positive and strongly negated RDF triples (so-called g-RDF triples) of the form $[\neg]p(s, o)$, where $p \in (V \cap URI) \cup Var$ is called property, and $s, o \in V \cup Var$ are the subject and the object, and a g-RDF program P containing derivation rules with possibly both explicit and scoped weak negation (Analyti et al. 2014). A g-RDF graph is a set of g-RDF triples. The g-RDF approach extends RDFS semantics and defines provenance stable models and provenance Herbrand interpretations. A g-RDF ontology O is a pair $\langle G_O, P_O \rangle$, where G_O is a g-RDF graph, and P_O is a g-RDF program, that is, a finite set of g-RDF rules, which are derivation rules of the form $t_0 \leftarrow t_1, \ldots, t_l, t_{l+1}@N_{l+1}, \ldots, t_k@N_k$, where each t_i for $i = 0, \ldots, k$) is a g-RDF triple and $N_i \subseteq URI$ for $i = l + 1, \ldots, k$. Automated reasoning over g-RDF graphs can utilize standard RDFS entailment and derivation rules of g-RDF ontologies.

RDFS is often used in combination with the *Web Ontology Language (OWL)*[7] to express complex relationships between classes, property cardinality constraints, characteristics of properties, equality of classes, enumerated classes, and so on (Sikos 2017b). In standard OWL, however, provenance is not covered. This is only available in extensions, such as OWL^C, which can define context-dependent axioms (Aljalbout et al. 2019).

1.4 Alternate Data Models and NoRDF Knowledge Representations

The provenance capturing limitations of RDF can be avoided by using alternate data models, although these typically do not have all the benefits of RDF, and simplicity in particular.

---

[6]https://www.w3.org/TR/rdf-schema/
[7]https://www.w3.org/OWL/

Knowledge organization systems that utilize RDFS and OWL represent entities as instances of classes, which facilitates deduction and integrity checking via automated reasoning (Sikos 2016). In contrast, if entities of knowledge bases are represented as embeddings in a vector space, new facts can be predicted via rule mining (using logic rules) or link prediction (using neural networks) (Suchanek et al. 2019), which cannot be done using RDF-based representations.

While RDF is an excellent choice for representing heterogeneous data, there are alternate data models for this purpose, such as *GSMM*, a graph-based meta-model purposefully designed for heterogeneous data management (Damiani et al. 2019). GSMM introduces nodes for subjects, predicates, and objects, and high-level constraints that must be satisfied. By modeling predicates as nodes, GSMM is backward-compatible with standard RDF reification.

1.5 RDF Graph Decomposition

Ding et al. (2005) proposed an intermediate (between document-level and triple-level) decomposition of RDF graphs into sets of *RDF molecules*, each of which is a connected subgraph of the original RDF graph. Formally, given an RDF graph G and a background ontology W, a decomposition $\widehat{G}$ of G is a set of RDF graphs $G_1, G_2, G_3, \ldots, G_n$, where G_i is a subgraph of G. The two operations related to this decomposition are (1) $\widehat{G} = d(G, W)$, which decomposes G to $\widehat{G}$ using W, and (2) $G' = m(\widehat{G}, W)$, which merges all subgraphs in $\widehat{G}$ into a new RDF graph R' using W. An RDF molecule m in G is a subgraph of G such that $m = d(m, W)$, where (d, m) represents pairs of lossless decompose-merge operations for which it holds that G is equivalent to $G' = m(d(G, W), W)$.

RDF molecules provide the right level of granularity suitable for tracking RDF provenance, evidence marshaling, and RDF document version control.

1.6 Encapsulating Provenance with RDF Triples

Using special-purpose triple elements in RDF statements for capturing provenance has the benefit of being compatible with standard triplestores, SPARQL queries, and RDF/RDFS entailment regimes, and enabling additional mechanisms for processing provenance data, even though these are proprietary.

The *Provenance Context Entity (PaCE)* approach was introduced as a viable alternative to RDF reification, with at least 49% less triples (Sahoo et al. 2010). It does not require the reification vocabulary nor blank nodes to trace RDF provenance. In contrast to RDF reification, PaCE has formal semantics defined using model theory and ensures backward-compatibility by extending the standard RDF and RDFS formal semantics. The PaCE approach does not define provenance at a fixed level; instead, it defines provenance at varying levels of granularity to suit application requirements. The heart of this approach is the *provenance context*, which is captured using formal objects representing a set of concepts and relationships required for a specific application to correctly interpret provenance-aware RDF statements. The provenance concept, pc, of an RDF triple (s, p, o) is a common object of the predicate `derives_from` associated with the triple. An RDFS-PaCE interpretation $\mathcal{I}$ of a vocabulary V is defined as an RDFS interpretation of the vocabulary $V \cup V_{PaCE}$ that satisfies the following extra condition: for RDF triples $\alpha = (S_1,P_1,O_1)$ and $\beta = (S_2,P_2,O_2)$, provenance-determined predicates (that are specified to the application domain), and entities v, if $pc(\alpha) = pc(\beta)$, then $(S_1,p,v) = (S_2,p,v)$, $(P_1,p,v) = (P_2,p,v)$, and $(O_1,p,v) = (O_2,p,v)$. A graph G_1 PaCE-entails a graph G_2 if every RDFS-PaCE interpretation that is a model of G_1 is also a model of G_2. Everything that can be inferred based on simple, RDF, or RDFS entailment is a PaCE entailment as well.

Another approach for making statements about other statements is using a *singleton property* specific to representing a particular fact according to a data source and further singleton properties to make other statements about the same RDF subject (Nguyen et al. 2014). In other words, each singleton property is unique to one particular entity pair. This is achieved by considering singleton properties as instances of generic properties, more specifically, by connecting singleton properties with their generic property using the `singletonPropertyOf` predicate. Provenance statements then can be made about the singleton properties to capture the data source, the extraction date, and so on.

The model-theoretic semantics of the singleton property approach is the extension of a simple interpretation I in a way that it satisfies these additional criteria: IP_S is a subset of IR, called the set of singleton properties of I, and $I_{S\_EXT}(p_s)$ is a function assigning to each singleton property a pair of entities from IR, formally $I_{S\_EXT}:IP_s \rightarrow IR \times IR$. As for an RDF interpretation $\mathcal{I}$, the semantics of the singleton property approach defines the following additional criteria: $x_s \in IP_s$ if $\langle x_s, \texttt{rdf:singletonPropertyOf}^{\mathcal{I}} \rangle \in I_{EXT}(\texttt{rdf:type}^{\mathcal{I}})$, $x_s \in IP_s$ if $\langle x_s,x^{\mathcal{I}} \rangle \in I_{EXT}(\texttt{rdf:singletonPropertyOf}^{\mathcal{I}})$, and $x \in IP$, $I_{S\_EXT}(x_s)= \langle s_1,s_2 \rangle$.

1.7 Capturing Context: Graph-Based Approaches

Based on the idea of *N-Quads*,[8] some approaches capture context for RDF triples using an additional tuple element, thereby forming RDF quadruples (quads). *Named graphs* are RDF graphs grouped with a unique identifier (graph name) about which additional statements can be made to capture context (Carroll et al. 2005). This approach triggered the development of other RDF reification alternatives, such as *RDF/S graphsets* (Pediaditis et al. 2009), which extends the RDFS semantics, and is not compatible with standard SPARQL queries. *Nanopublications* are the smallest

---

[8]https://www.w3.org/TR/n-quads/

units of publishable information, consisting of three elements (typically represented as named graphs): assertion, provenance, and publication metadata (Groth et al. 2010). *RDF triple coloring* (Flouris et al. 2009) uses "colors" to capture data integration scenarios in which the same data was derived from different resources. *Hoganification* is a graph-based provenance capturing technique that combines the benefits of some of the previous approaches (Hogan 2018).

Although purpose-designed datastores are available for storing and querying RDF quadruples (which are called quadstores), there are no general reasoning rules for RDF quads because what RDF quadruples represent are arbitrary, which prevents context-aware reasoning over RDF quads. Therefore, quad-based provenance capturing is limited when it comes to reasoning over data provenance and not just the data itself.

1.8   Utilizing Vocabularies and Ontologies

There are approaches for capturing RDF provenance that define a knowledge organization system as the core of their mechanism or implementation technique. The aforementioned PaCE approach, for example, relies on an external ontology called the *Provenir* ontology (Sahoo and Sheth 2009). The singleton property utilizes a single predicate, `singletonPropertyOf`, on top of standard RDF and RDFS predicates, to capture provenance. *NdFluents* creates contextualized individuals for both the subject and the object of RDF triples using a purpose-built ontology that defines `contextualPartof` and `contextualExtent` predicates (Giménez-García et al. 2017). The triples are then replaced by a new triple that uses the contextualized entities. The two new resources are related to the original individuals and with a context to which the annotations are attached. Formally, an original RDF triple (s, p, o) is replaced by the set of triples $\{(s_c, p, o_c), (s_c, \text{nd:contextualPartOf}, s), (o_c, \text{nd:contextualPartOf}, o), (s_c, \text{nd:contextualExtent}, c), (o_c, \text{nd:contextualExtent}, c)\}$, where c is a function of the context and annotations are related to c.

An application-specific example for ontology-based RDF provenance capturing is the *Communication Network Topology and Forwarding Ontology (CNTFO),*[9] which can be used in the networking and cybersecurity domains (Sikos et al. 2018).

It was designed as part of a named graph-based approach, *GraphSource*, to capture provenance, especially for cyber-situational awareness applications (Philp et al. 2019). CNTFO defines communication network concepts from the routing perspective in a taxonomical structure. The types of concepts in this ontology include network topology concepts, network events, logical groupings, and networking technologies. The properties of these concepts are also defined so that connections between devices can be described formally and accurately. CNTFO

---

[9]http://purl.org/ontology/network/

defines background knowledge of the network domain, and by instantiating the classes of the ontology, network paths and routing messages can be described. The datatype declarations of the ontology can be used to verify/restrict permissible property value ranges for the properties of network concepts. CNTFO is aligned with the industry standard provenance ontology, *PROV-O*.[10] It provides seamless integration with named RDF graphs for capturing statement-level provenance, and with the *Vocabulary of Interlinked Datasets (VoID)*[11] for dataset-level provenance.

1.9 Capturing Metadata with or Without Provenance

There are approaches developed for capturing other types of metadata, such as temporal constraints, for RDF. While not designed primarily or exclusively for provenance representation, some of these may also be suitable for capturing provenance, such as *temporal RDF* (Hurtado and Vaisman 2006; Gutierrez et al. 2007; Tappolet and Bernstein 2009).

Generally speaking, temporal RDF graphs are sets of temporal RDF triples of the form $(s, p, o) : [\tau]$, where $\tau \in \mathbb{N}_0$ is a temporal label, where 0 is the lowest possible value referring to the beginning of the considered time period, modeling time as a one-dimensional discrete value (not considering branching time). τ is a time interval ranging from start time s to end time e such that $s, e \in \tau | s \leqslant e$. The exact implementation depends on the approach, but they all sit on top of the RDF data model and need extensions to SPARQL to query time points and intervals.

1.10 Summary

The purity of the standard RDF data model has several benefits, but the incapability of capturing data provenance makes RDF unusable in application fields for which provenance is of utmost importance for verification, tracking, and other purposes. The various approaches and techniques proposed over the years to capture RDF provenance all have their strengths and weaknesses; therefore, which one to choose depends on the application. Generally, graph-based approaches outperform instantiated predicate-based approaches in terms of querying, although reasoning over RDF provenance has additional requirements beyond capturing provenance for individual RDF statements of sets of RDF triples. It is important to keep in mind that even those mechanisms that have formal semantics defined not necessarily have semantics for every component, such as unique predicates with consecutive numbering for different relationship versions depending on the data source are just known to be derivatives of the original predicate. Not only that, but part of the

---

[10]http://www.w3.org/ns/prov-o
[11]https://www.w3.org/TR/void/

semantics of these might only make sense to humans and cannot be fully interpreted by software agents alone.

References

Aljalbout S, Buchs D, Falquet G (2019) Introducing contextual reasoning to the Semantic Web with OWL^C. In: Endres D, Alam M, Şotropa D (eds) Graph-based representation and reasoning. Springer, Cham, pp 13–26. https://doi.org/10.1007/978-3-030-23182-8_2

Analyti A, Damásio CV, Antoniou G, Pachoulakis I (2014) Why-provenance information for RDF, rules, and negation. Ann Math Artif Intel 70(3):221–277. https://doi.org/10.1007/s10472-013-9396-0

Berners-Lee T, Connolly D, Kagal L, Scharf Y, Hendler J (2008) N3Logic: a logical framework for the World Wide Web. Theory Pract Log Program 8(3):249–269. https://doi.org/10.1017/S1471068407003213

Carroll JJ, Bizer C, Hayes P, Stickler P (2005) Named graphs, provenance and trust. In: Proceedings of the 14th International Conference on World Wide Web. ACM, New York, pp 613–622. https://doi.org/10.1145/1060745.1060835

Damiani E, Oliboni B, Quintarelli E, Tanca L (2019) A graph-based meta-model for heterogeneous data management. Knowl Inf Syst 61(1):107–136. https://doi.org/10.1007/s10115-018-1305-8

Ding L, Finin T, Peng Y, da Silva PP, McGuinness DL (2005) Tracking RDF graph provenance using RDF molecules. In: Fourth International Semantic Web Conference

Dividino R, Sizov S, Staab S, Schueler B (2009) Querying for provenance, trust, uncertainty and other meta knowledge in RDF. J Web Semant 7(3):204–219. https://doi.org/10.1016/j.websem.2009.07.004

Flouris G, Fundulaki I, Pediaditis P, Theoharis Y, Christophides V (2009) Coloring RDF triples to capture provenance. In: Bernstein A, Karger DR, Heath T, Feigenbaum L, Maynard D, Motta E, Thirunarayan K (eds) The Semantic Web – ISWC 2009. Springer, Heidelberg, pp 196–212. https://doi.org/10.1007/978-3-642-04930-9_13

Giménez-García JM, Zimmermann A, Maret P (2017) NdFluents: an ontology for annotated statements with inference preservation. In: Blomqvist E, Maynard D, Gangemi A, Hoekstra R, Hitzler P, Hartig O (eds) The Semantic Web. Springer, Cham, pp 638–654. https://doi.org/10.1007/978-3-319-58068-5_39

Groth P, Gibson A, Velterop J (2010) The anatomy of a nanopublication. Inform Serv Use 30(1–2):51–56. https://doi.org/10.3233/ISU-2010-0613

Gutierrez C, Hurtado CA, Vaisman A (2007) Introducing time into RDF. IEEE T Knowl Data En 19(2):207–218. https://doi.org/10.1109/TKDE.2007.34

Hartig O, Thompson B (2014) Foundations of an alternative approach to reification in RDF. https://arxiv.org/abs/1406.3399

Hartig O, Thompson B (2019) Foundations of an alternative approach to reification in RDF. https://arxiv.org/abs/1406.3399. arXiv:1406.3399

Hoffart J, Suchanek FM, Berberich K, Weikum G (2013) YAGO2: a spatially and temporally enhanced knowledge base from Wikipedia. Artif Intell 194:28–61. https://doi.org/10.1016/j.artint.2012.06.001

Hogan A (2018) Context in graphs. In: Proceedings of the 1st International Workshop on Conceptualized Knowledge Graphs. RWTH Aachen University, Aachen

Hurtado C, Vaisman A (2006) Reasoning with temporal constraints in RDF. In: Alferes JJ, Bailey J, May W, Schwertel U (eds) Principles and practice of semantic web reasoning. Springer, Heidelberg, pp 164–178. https://doi.org/10.1007/11853107_12

Keskisärkkä R, Blomqvist E, Lind L, Hartig O (2019) RSP-QL*: enabling statement-level annotations in RDF streams. In: Semantic systems. The power of AI and knowledge graphs. Springer, Cham, pp 140–155. https://doi.org/10.1007/978-3-030-33220-4_11

Nguyen V, Bodenreider O, Sheth A (2014) Don't like RDF reification?: Making statements about statements using singleton property. In: Proceedings of the 23rd International Conference on World Wide Web. ACM, New York, pp 759–770. https://doi.org/10.1145/2566486.2567973

Pediaditis P, Flouris G, Fundulaki I, Christophides V (2009) On explicit provenance management in RDF/S graphs. In: First Workshop on the Theory and Practice of Provenance, San Francisco, CA, USA, 23 February 2009

Philp D, Chan N, Sikos LF (2019) Decision support for network path estimation via automated reasoning. In: Czarnowski I, Howlett RJ, Jain LC (eds) Intelligent decision technologies 2019. Springer, Singapore, pp 335–344. https://doi.org/10.1007/978-981-13-8311-3_29

Sahoo SS, Sheth A (2009) Provenir ontology: towards a framework for eScience provenance management. Microsoft eScience Workshop

Sahoo SS, Bodenreider O, Hitzler P, Sheth A, Thirunarayan K (2010) Provenance Context Entity (PaCE): scalable provenance tracking for scientific RDF data. In: Gertz M, Ludäscher B (eds) Scientific and statistical database management. Springer, Heidelberg, pp 461–470. https://doi.org/10.1007/978-3-642-13818-8_32

Sikos LF (2016) A novel approach to multimedia ontology engineering for automated reasoning over audiovisual LOD datasets. In: Nguyen NT, Trawiski B, Fujita H, Hong TP (eds) Intelligent information and database systems. Springer, Heidelberg, pp 3–12. https://doi.org/10.1007/978-3-662-49381-6_1

Sikos LF (2017a) 3D model indexing in videos for content-based retrieval via X3D-based semantic enrichment and automated reasoning. In: Proceedings of the 22nd International Conference on 3D Web Technology. ACM, New York. https://doi.org/10.1145/3055624.3075943

Sikos LF (2017b) Description logics in multimedia reasoning. Springer, Cham. https://doi.org/10.1007/978-3-319-54066-5

Sikos LF (2017c) A novel ontology for 3D semantics: ontology-based 3D model indexing and content-based video retrieval applied to the medical domain. Int J Metadata Semant Ontol 12(1):59–70. https://doi.org/10.1504/IJMSO.2017.087702

Sikos LF (2018a) Ontology-based structured video annotation for content-based video retrieval via spatiotemporal reasoning. In: Kwaśnicka H, Jain LC (eds) Bridging the semantic gap in image and video analysis. Springer, Cham, pp 97–122. https://doi.org/10.1007/978-3-319-73891-8_6

Sikos LF (2018b) Spatiotemporal reasoning for complex video event recognition in content-based video retrieval. In: Hassanien AE, Shaalan K, Gaber T, Tolba MF (eds) Proceedings of the International Conference on Advanced Intelligent Systems and Informatics 2017. Springer, Cham, pp 704–713. https://doi.org/10.1007/978-3-319-64861-3

Sikos LF, Philp D (2020) Provenance-aware knowledge representation: a survey of data models and contextualized knowledge graphs. Data Sci Eng. https://doi.org/10.1007/s41019-020-00118-0

Sikos LF, Stumptner M, Mayer W, Howard C, Voigt S, Philp D (2018) Representing network knowledge using provenance-aware formalisms for cyber-situational awareness. Procedia Comput Sci 126C:29–38. https://doi.org/10.1016/j.procs.2018.07.206

Suchanek FM, Lajus J, Boschin A, Weikum G (2019) Knowledge representation and rule mining in entity-centric knowledge bases. In: Krötzsch M, Stepanova D (eds) Reasoning Web. Explainable artificial intelligence, chap 4. Springer, Cham, pp 110–152. https://doi.org/10.1007/978-3-030-31423-1_4

Tappolet J, Bernstein A (2009) Applied temporal RDF: efficient temporal querying of RDF data with SPARQL. In: Aroyo L, Traverso P, Ciravegna F, Cimiano P, Heath T, Hyvönen E, Mizoguchi R, Oren E, Sabou M, Simperl E (eds) The Semantic Web: research and applications. Springer, Heidelberg, pp 308–322. https://doi.org/10.1007/978-3-642-02121-3_25

Zimmermann A, Lopes N, Polleres A, Straccia U (2011) A general framework for representing, reasoning and querying with annotated Semantic Web data. Web Semant Sci Serv Agents World Wide Web 11:72–95. https://doi.org/10.1016/j.websem.2011.08.006

Chapter 2
Data Provenance and Accountability on the Web

Oshani W. Seneviratne (iD)

2.1 Data Longevity

Thanks to the many content dissemination techniques, data on the Web can take a life of its own. For example, one can copy a web page and create another page on a different server. Many content aggregation services collect resources, and on social networking platforms, such as Twitter and Facebook, users re-(tweet/share/post) other users' content. The longevity of data can have good (Sect. 2.1.1), bad (Sect. 2.1.2), and ugly (Sect. 2.1.3) implications.

2.1.1 The Good

There is tremendous good to society when users share the right information with the right people in the right ways. This section outlines several domains in which data reuse with proper controls leads to good outcomes.

Scientists can use data in unexpected ways and discover groundbreaking results in life sciences (Marshall et al. 2012) and find treatments for diseases, as evidenced by the COVID-19 pandemic, during which some organizations have started disseminating data in RDF (Ilievski et al. 2020). Furthermore, there is no surprise that data and code sharing in the research world can lead to further discoveries. Data sharing and subsequent secondary research have received some attention since the infamous "research parasites" (Longo and Drazen 2016). Swiftly condemned by many researchers in statistics and data science (Oransky and Marcus 2016), it has resulted in the establishment of a "Research Parasite Award" (Greene Laboratory

O. W. Seneviratne (✉)
Health Data Research, Rensselaer Polytechnic Institute, Troy, NY, USA
e-mail: senevo@rpi.edu

© Springer Nature Switzerland AG 2021
L. F. Sikos et al. (eds.), *Provenance in Data Science*, Advanced Information and Knowledge Processing, https://doi.org/10.1007/978-3-030-67681-0_2

2017) for excellence in secondary data analysis. While the award has a derogatory title, it is clear that data sharing has made a difference to scientific advancement. However, since a significant investment amount has already gone into the original dataset creation, the original dataset creator needs to track how their data has been utilized in subsequent analyses.

Furthermore, in the world of user-generated creative content, we see that the Web's remix culture promotes people to build off on each other's work (Seneviratne and Monroy-Hernandez 2010). The Web empowers the creation of knowledge and culture by enabling a medium in which content creators can reuse previous works. Manovich points out that "remixing is practically a built-in feature of the digital networked media universe" (Manovich 2005). Benkler posits this as a fundamental part of the way culture is produced and argues that if "we are to make this culture our own, render it legible, and make it into a new platform for our needs and conversations today, we must find a way to cut, paste, and remix present culture" (Benkler 2006). Many of the existing websites that promote reuse could use some design interventions such as letting content creators choose the license, such as a Creative Commons (CC) license (Lessig 2003), for their work, and display it in a human-readable, easy-to-understand short text rather than in verbose license terms only an expert such as a lawyer would understand. Encoding the license terms in a machine-interpretable format, such as RDF, allows user-agents such as web browsers to display the license terms appropriate to the target audience.

2.1.2 The Bad

Many naïve users may have the impression that the website to which they disclose their data will not misuse them. They may not realize that when their personal information is taken out of context and reused in a way that was not initially intended, there may be unseen adverse consequences for them. For example, in the healthcare and finance domains, the data used is reasonably complex and unpredictable. The user may not be completely aware of what is happening with their information, and the potential privacy implications of the information mergers and misuses.

The famous phrase "if the product is free, you are the product" exemplifies many companies' penchant to offer information and services of any kind on the Web in exchange for users' data. Users who sign up for such services disclosing their data may often find themselves receiving unsolicited promotional offers. Even worse, they may realize later that their data have been sold to another party. One such example is the data leakage from Facebook, in which 50 million users were profiled and targeted without their consent by Cambridge Analytica, a third-party political analytics firm (Cadwalladr and Graham-Harrison 2018).

Personal privacy management is challenging on social media websites because it is effortless to create shadow profiles on which most of the information about a certain user is available via other peoples' public posts (Garcia 2017). For example, one user may tweet "happy birthday @ProtectedUser," publicly creating a shadow record of the protected user's birthday. Then another public account tweets "looking forward to our trip," again exposing data or creating a tag for the protected user being together with the referred person during some event at a particular location (Keküllüoglu et al. 2020). Together these public posts create a shadow profile of the protected user, which cannot be easily controlled.

Furthermore, people often post their creative content, assuming that it will be quoted, copied, and reused. They may wish that their work will be subject to restrictions that they prefer, not the restrictions of the websites they post their content to. Such user-preferred restrictions may include that their content will only be used with attribution, used only for non-commercial use (or used for commercial use they will get compensated), or distributed under a license similar to that of the original content, and so on. Attributing one's work is a form of social compensation and an attention mechanism to publicize one's creative work. In an experiment conducted on Flickr image attribution, it was discovered that up to 94% of the reused images might not be properly attributing the original owner of the image, even though the license may be explicitly specifying to do so (Seneviratne et al. 2009).

2.1.3 The Ugly

Online social networks are used by billions to connect with each other and share significant life milestones or stressful life events for emotional support (Andalibi and Forte 2018; Yang et al. 2019). However, sharing life events can lead to unintended disclosures, mainly because social network users often do not know how to estimate the audience size of their posts (Bernstein et al. 2013). Analyzing life events on Twitter and the privacy leaks have been the subject of Keküllüoğlu et al. (2020), in which Twitter users' key life events have been automatically identified. This indicates that even if a user has a protected account, there are potential implications for privacy, particularly around the impact of connections' sharing decisions. Even if a user is privacy-conscious and can accurately estimate their posts' reach, their networks could still disclose them. For example, users can share their location on an online social network and tag one another, inherently revealing where they are. As we witness in many news reports and studies (Keküllüoglu et al. 2020), naïve users do not have a clear understanding of their online privacy, and the data disclosed in social networking profiles is often used against them (Duhigg 2012).

The highly public nature of platforms such as Twitter also means that information shared is open to a worldwide audience. Users are not completely safe even when they restrict the posts' audience because replies to the original post may lead to privacy leaks (Jain et al. 2013). A post about selling an item, including the seller's phone number, might be retweeted or sent on by well-meaning followers. Furthermore, peculiar design decisions employed by some online social media platforms may further complicate things. For example, interactions with a tweet increase its visibility, and even if the original tweet is protected or deleted, public replies to it will stay on the platform. These replies might signal the events happening in the original tweet, which cannot be managed by the event subject. Users may know that there could be potential privacy leaks, but it is probably hard for them to conceptualize that there are potential adverse consequences, mainly because many believe that they are not interesting enough to be targets for attackers (Wash 2010). Personal attributes such as age, gender, religion, diet, and personality can be inferred using the tweets a user is mentioned in Jurgens et al. (2017). Also, analyzing a user's connection network might suggest homophily and latent attributes about them, such as age group, gender, or sexual orientation (Jernigan and Mistree 2009), location or location affinity (Zamal et al. 2012), and political orientation (Aldayel and Magdy 2019). Some data exposures result in minor discomfort, such as colleagues learning that the user is moving, leaving their job, or getting a divorce. However, some posts can cause severe damage to someone's financial affairs and career prospects. For example, posts about surgeries or illnesses can result in insurance premium increases or the loss of their job (Heussner 2009).

2.2   Data Provenance

It is essential to understand that data is a living, breathing ecosystem that needs to be continually classified, tracked, and cataloged. Classification is much manageable when an algorithm can help us decide what things like personally identifiable information or creative content are, and we take appropriate steps to anonymize data or provide a mechanism to share data appropriately. When ownership is attributed between individuals and groups of individuals, it takes longer to grant access. Consent only works if rules are followed, which means that appropriate systems to track data lineage need to be in place, thereby ensuring that users do what they say they are doing with the data they have been consented to use. Therefore, data would need to be tracked, and copies of the data should be acknowledged. It should be understood where this data is copied to and in what form. There are critical concerns like the right to be forgotten, which is enshrined in legislation such as the *General Data Protection Regulation (GDPR)* (Voigt and von dem Bussche 2015). Such aspects should be front and center of a data representation strategy. A data strategy should, therefore, encompass data management that enables data to be consented and tracked. The level to which data access and reuse can be audited ensures that the correct processes are in place should a breach occur. How we collect data and the rate it is collected at is an essential factor as well. Sometimes it is not appropriate to de-identify data, so it is mandatory to understand who is processing your data. We will look at some of the techniques for representing the terms associated with data in a machine-processable manner.

2.2.1 Rights Expression

Data reuse can be simplified by providing the necessary tools that leverage machine-readable rights expressed in RDF and displayed in human-friendly mechanisms to make the users more aware of the usage restriction options available and ensure that the user is compliant.

García et al. (2013) and Finin et al. (2008) offer significant contributions to how end-to-end usage rights and access control systems may be implemented in the Web Ontology Language (OWL) and RDF. Garcìa proposes a "Copyright Ontology" based on OWL and RDF for expressing rights, representations that can be associated with media fragments in a web-scale "rights value change." Finin describes two ways to support standard role-based access control models in OWL, and discusses how their OWL implementations can be extended to model attribute-based access control.

To allow use restrictions on creative content, Creative Commons offers four distinct license types: BY (attribution), NC (non-commercial), ND (no derivatives), and SA (share-alike) that can be used in combination to best reflect the content creator's rights. To facilitate such varied application of usage restrictions, users have the flexibility to express their rights utilizing the *Creative Commons Rights Expression Language (ccREL)*.[1] ccREL empowers the users by not forcing them into choosing a predefined license for their works or having to agree with the website's terms without any flexibility at all, as is the case with many user-generated content websites. ccREL also allows a publisher of work to give additional permissions beyond those specified in the CC license with the use of the `cc:morePermissions` property embedded in RDF in attributes (RDFa) that the user can utilize to reference commercial licensing brokers or any other license deed and `dc:source` to reference parent works.

Various technologies allow one to transfer metadata along with the content by embedding the metadata in machine-interpretable RDF. *Extensible Metadata Platform (XMP)* (Ball 2007) is widely deployed for embedding usage restrictions in multimedia content such as images, audio, and video. Another format that is nearly universal when it comes to images is the *Exchangeable Image File Format (EXIF)* (JEITA 2002). The metadata tags available in these standards cover a broad spectrum, including date & time information, camera settings, thumbnail, and, more importantly, the description of the photos, including copyright information.

A document fragment ontology representing information about the sources of its content and a technique for extracting from these documents that allow programs to trace the excerpt back to the source is introduced by Jones (2007). In a similar vein, the *Semantic Clipboard*[2] gleans the Creative Commons license information expressed in RDFa from HTML pages. Using this tool, as the user browses several

---

[1] https://www.w3.org/Submission/ccREL/
[2] http://dig.csail.mit.edu/2009/Clipboard.

menu options in the Firefox browser can be used to select licensed images with the proper intention, for example, commercial purpose. Finally, when a user attempts to copy an image from the browser, the tool will automatically determine if it is permissible by checking the license restrictions. If it is permitted, it will create the appropriate license.

2.2.2 Provenance System Implementations

Provenance is defined as a record describing how organizations, people, data, and activities may have influenced a decision or an outcome of a process. Provenance can be employed to track which organization, people, sensors, execution of a previous plan may have influenced the domain knowledge used in the planning process and, eventually, the resulting plan and its likelihood to succeed. Recording the provenance of a plan and plan execution provides us with an audit trail to trace back all the influences that went into the plan's generation and what may have impacted its implementation.

Provenance has been traditionally used to prove lineage of data in scientific computing (Simmhan et al. 2005), scientific workflows (Davidson and Freire 2008), on time-stamping a digital document (Haber and Stornetta 1991), and, more recently, to provide metadata about the origin of data (Moreau et al. 2008). The principles for documentation of proper evidence of past execution, as stated by Groth et al. (2012), include the following: (i) "assertions must be based on observations made by software components, according to the principle of data locality," (ii) "each assertion making up documentation of the past for a computer system must be attributable to a particular software component." Using the above principles, provenance management tools such as *ProvToolBox* have created Java representations of the Prov data model and manipulate it from the Java programming language (Magliacane 2012). ProvenanceJS is a Javascript tool that can embed and extract provenance from HTML pages (Groth 2010).

Inference Web (McGuinness and Da Silva 2004) was designed to be an extension of the Semantic Web that aims to enable applications to generate explanations for any of their answers supported by the *Proof Markup Language (PML)* (Da Silva et al. 2006). PML includes support for knowledge provenance, reasoning information, metadata such as authorship and authoritativeness of a source, reasoner assumptions (e.g., closed vs. the open world, naming assumptions) detailed trace of inference rules applied. Human-readable explanations can be derived from PML and browsable representations can also be displayed to the user.

2.2.3   Limitations of Provenance

While provenance is useful in many aspects, it may not be adequate by itself in system implementations due to several reasons. This is primarily due to the problems arising from the breakage of provenance and techniques capturing and preserving the provenance of a derivative work created from more than one disparate knowledge source. Even if provenance metadata was created correctly in RDF, they might be easily separated from the data or the content that the provenance information is applied to, thus breaking the chain of provenance. Similarly, for interoperability, when two or more datasets or contents are reused together, the user needs to ascertain what permissions the newly created dataset would inherit, and there may be some incompatibilities when combining the provenance representations, especially if there are incompatible schemata for data representation.

2.3   Data Accountability

Accountable systems make both data provenance and transparency, and reuse first-class citizens, enabling the information owner to check how their resources have been used. The usage data, expressed in RDF, can be reasoned with usage restrictions defined by the content creator or any other policy representation, expressed in RDF, to assert that no violation has occurred. One of the main goals in accountable systems is that users are aware of and control their data on the Web. Another goal of accountable systems is to let content contributors see how their content has been used and enable content consumers to use them appropriately. This objective is vastly different from access control systems, which have been the de facto privacy, rights, and data protection mechanisms in many systems. In a pure access restriction system, those who obtain access to the data, legitimately or not, would use the data without any restriction due to the content's digital nature. In addition, there may be situations where restricting access could cause more harm than good. Having unfettered access to the right information at the right time by the right people, in a situation such as a healthcare emergency, regardless of them being authorized to access that information in the first place, might make the difference between life and death. Furthermore, insider threats can be a substantial issue that motivates the need for data usage accountability. Most recently, many reported privacy breaches happened through authorized insiders' work accessing and using private information in inappropriate ways, and not by malicious attackers from the outside.

2.3.1   Definition of Accountable Systems

In some early work on accountability in computer science, deficiencies in protection mechanisms have been identified with solutions that put constraints on the use of information after releasing by one program (Saltzer and Schroeder 1975). The Halpern-Pearl framework of causality (Halpern and Pearl 2005a,b) gives formal definitions of both responsibility and blame to inform actions and punishment taken in response to a policy violation. Küsters et al. (2010) have proposed a model-independent, formal definition of accountability focused on systems to provide substantial evidence of misbehavior, where it has been applied in three cryptographic tasks, including contract-signing, voting, and auctions.

Formalization of accountability includes the unified treatment of scenarios in which accountability is enforced automatically and executed by a mediating authority (Feigenbaum et al. 2011b). Another aspect is to handle scenarios in which those held accountable may remain anonymous but must be identifiable when needed (Feigenbaum et al. 2011a). The three broad aspects of accountability (time, information, and action) have also been explored from a theoretical lens (Feigenbaum et al. 2012).

Characteristics and the mechanism for realizing accountable systems are outlined in Seneviratne (2014), which focused on technologies that enable appropriate use of data and help determine inappropriate use of data through provenance. In such a system, information can be accessed whenever necessary. However, it must log all activity in a privacy-preserving manner for compliance checks afterward to determine inappropriate uses of information. Subsequently, enabling tracking content use is the ideal compromise between prohibitively restrictive access control mechanisms and a free for all "Wild West" of dataspaces.

2.3.2   Implementations of Accountable Systems

The PeerReview System (Haeberlen et al. 2007) has implemented the notion of accountability in distributed systems. The system maintains a tamper-evident record that provides non-repudiable evidence of all participants' actions and combines that with a tamper-evident log that each node maintains about its interactions, and can be obtained by other nodes on demand. Work on distributed usage control (Pretschner et al. 2006; Kumari et al. 2011) proposes enforcement of usage restrictions regarding privacy-sensitive personal data at various levels of abstractions in the operating system. Such distributed usage control policies stipulate rights and duties concerning the future use of data using a data flow tracking within and across layers of abstraction as well as within and across single machines, declassification based on data quantities, provenance tracking, and ways to protect the infrastructure itself.

2.3.3   Ensuring Accountability on the Web

The Web empowers the creation of knowledge and culture through creative works, and it is one of the primary media using which people share personal information and engage with each other. This section will examine some architecture for decentralized systems, such as the Web, that utilize principles of RDF and provenance.

As noted earlier, provenance is a crucial enabler for accountable web systems. Such systems comprise an explicit representation of past processes that allows us to trace the origin of data, actions, and decisions by automated or human processes. Special-purpose protocols have demonstrated to augment the Web with accountability (Seneviratne 2012) both for private data sharing and creative work dissemination. The HyperText Transfer Protocol with Accountability (HTTPA) is an end-to-end protocol that extends to provide appropriate use negotiation for accountability (Seneviratne 2014), which requires both servers and user-agents all to agree on the protocol terms. In HTTPA, when a user-agent requests a resource, the server will respond with that resource's usage restrictions expressed in RDF. For example, the usage restriction could be "do not share this resource," signaling that this resource is only intended for the requestor's eyes. The client will send the agreement to the usage restriction in a special HTTPA request. Then the server sends the transaction information to be logged in a special-purpose network of servers called the Provenance Tracking Network (PTN). Logs in HTTPA are immutable except by protocol components, readable only by the party involved in the HTTPA transaction, and have all the pertinent records of a particular data transfer and use. The logs are annotated using the *World Wide Web Consortium (W3C) Provenance Ontology (PROV-O)*[3] and are stored as named graphs (Carroll et al. 2005) associated with the resource in question. Once the usage restrictions are agreed upon, and the transaction is logged in the PTN, the server will send the resource to the client. There is no strict enforcement of resource use, only an end-to-end mechanism to keep the users accountable. The resource owner can consult the PTN to request a compliance log for that resource to ascertain no usage restriction violations have occurred. The PTN only stores pointers to the resource and the data related to the access, repurpose, and share activities. The definition of HTTPA usage restrictions is up to the developers, as long as they use a machine-processable format (such as RDF) to describe the terms (Senevitane and Kagal 2011). Similarly, a privacy auditing framework that utilizes ontologies that supports provenance-enabled logging of events and synthesized contextual integrity for compliance and observation derivation is described by Samavi and Consens (2012).

When deployed on the Web, such systems may provide an attractive compromise with parties' rights and abilities to control access and use are balanced against the burden of the imposed restrictions. Even though it may sound like an oxymoron,

---

[3]https://www.w3.org/TR/prov-o/.

there may be a mechanism that enables privacy through the transparency of user actions on the resource in question, as described by Seneviratne and Kagal (2014).

2.4   Future Directions

One of the driving factors behind the current Artificial Intelligence (AI) revolution is the massive amount of data available for research, development, and commerce. Data is sometimes referred to as the "new gold" or "new currency." However, there are high barriers to getting access to this gold. It is challenging to obtain sufficient data for training models, unless the said research is conducted in one of the large AI powerhouses with access to Big Data. To address this issue, accountable systems powered by RDF technologies can enable a marketplace for data for the AI models, thereby making access to data more democratic. Dataset owners and AI model creators can be rewarded for their contributions, the authenticity of the data, and other criteria such as the quality of the AI models' results.

An alternative technique for implementing accountable systems is through Distributed Ledger Technologies (DLT). DLT could be used to preserve the provenance of data records cryptographically, making it impossible to deny any misuse of data. Coupled with a robust and expressive mechanism such as RDF, this may result in the next generation of accountable system implementations. For example, some of the aspects described for accountable systems would be automated using smart contracts (Buterin 2014) deployed on DLTs both for the model execution and the reward mechanism.

2.5   Conclusion

Tools and standards we use every day must strike an appropriate balance between honoring individuals' rights and the power of reuse. Access control systems are often successful in managing access to resources, but they provide very little help to users to trace how their content has been shared and used by others, particularly with personal data and creative content. With personal data, there is an incentive for users to share their data, for example health data, with key personnel (such as healthcare providers) if there is a "break-glass-in-case-of-emergency" situation. Similarly, with creative content, there is an inherent sense of pride when others use one's work, even though there is no direct financial incentive. Enabling and determining the appropriate use of information on the Web concerning privacy and intellectual property rights violations is an essential consideration for the Web to realize its full potential.

Expressing what constitutes appropriate use of a resource and restrictions on those resources using a standards-compliant rights expression language based on RDF is an excellent way to preserve the rights held on a resource and trace its reuse.

By providing the necessary tools to leverage such machine-readable rights and making users aware of the available usage restriction options, we will ensure that they will comply with the simplified mechanisms when reusing data and resources available at their disposal.

As a consequence, tools and techniques are needed that can be used to enable the appropriate use of information by making it easy for the user to do the right thing. Accountable systems inherently manage the provenance of their contents and have the flexibility to capture any level of granularity of information required. By using existing recommended standards like RDF, OWL, and SPARQL, accountable systems augmented with provenance, such as HTTPA (Seneviratne 2014), can provide flexible expansion and integration options without the limitations of custom database management tools.

References

Aldayel A, Magdy W (2019) Your stance is exposed! analyzing possible factors for stance detection on social media. Proc ACM Hum-Comput Interact 3. https://doi.org/10.1145/3359307

Andalibi N, Forte A (2018) Announcing pregnancy loss on Facebook: A decision-making framework for stigmatized disclosures on identified social network sites. In: Proceedings of the 2018 CHI conference on human factors in computing systems. https://doi.org/10.1145/3173574.3173732

Ball A (2007) Briefing paper: The Adobe eXtensible Metadata Platform (XMP). UKOLN Research Organization. http://www.ukoln.ac.uk/projects/grand-challenge/papers/xmpBriefing.pdf

Benkler Y (2006) The wealth of networks: how social production transforms markets and freedom. Yale University Press, London

Bernstein MS, Bakshy E, Burke M, Karrer B (2013) Quantifying the invisible audience in social networks. In: Proceedings of the SIGCHI conference on human factors in computing systems, pp 21–30. https://doi.org/10.1145/2470654.2470658

Buterin V (2014) A next-generation smart contract and decentralized application platform. White Paper. https://blockchainlab.com/pdf/Ethereum_white_paper-a_next_generation_smart_contract_and_decentralized_application_platform-vitalik-buterin.pdf

Cadwalladr C, Graham-Harrison E (2018) Revealed: 50 million Facebook profiles harvested for Cambridge Analytica in major data breach. Guardian 17:22. https://www.theguardian.com/news/2018/mar/17/cambridge-analytica-facebook-influence-us-election

Carroll JJ, Bizer C, Hayes P, Stickler P (2005) Named graphs. J Web Semant 3(4):247–267. https://doi.org/10.1016/j.websem.2005.09.001

Da Silva PP, McGuinness DL, Fikes R (2006) A proof markup language for Semantic Web services. Inf Syst 31(4–5):381–395. https://doi.org/10.1016/j.is.2005.02.003

Davidson SB, Freire J (2008) Provenance and scientific workflows: challenges and opportunities. In: Proceedings of the 2008 ACM SIGMOD international conference on management of data, pp 1345–1350. https://doi.org/10.1145/1376616.1376772

Duhigg C (2012) How companies learn your secrets. The New York Times. https://www.nytimes.com/2012/02/19/magazine/shopping-habits.html

Feigenbaum J, Hendler JA, Jaggard AD, Weitzner DJ, Wright RN (2011a) Accountability and deterrence in online life. In: Proceedings of the 3rd international web science conference, pp 1–7. https://doi.org/10.1145/2527031.2527043

Feigenbaum J, Jaggard AD, Wright RN (2011b) Towards a formal model of accountability. In: Proceedings of the 2011 new security paradigms workshop, pp 45–56. https://doi.org/10.1145/2073276.2073282

Feigenbaum J, Jaggard AD, Wright RN, Xiao H (2012) Systematizing "accountability" in computer science. Tech. rep. http://www.cs.yale.edu/publications/techreports/tr1452.pdf, technical Report YALEU/DCS/TR-1452, Yale University, New Haven, CT

Finin T, Joshi A, Kagal L, Niu J, Sandhu R, Winsborough W, Thuraisingham B (2008) ROWLBAC: representing role-based access control in OWL. Proc ACM Symp Access Control Models Technol, pp 73–82. https://doi.org/10.1145/1377836.1377849

Garcia D (2017) Leaking privacy and shadow profiles in online social networks. Sci Adv 3(8):e1701172

García R, Castellà D, Gil R (2013) Semantic copyright management of media fragments. In: Proceedings of the 2nd international conference on data technologies and applications, pp 230–237

Greene Laboratory (2017) The Parasite Awards: celebrating rigorous secondary data analysis. https://researchparasite.com

Groth P (2010) ProvenanceJS: revealing the provenance of web pages. In: McGuinness DL, Michaelis JR, Moreau L (eds) Provenance and annotation of data and processes. Springer, Heidelberg, pp 283–285. https://doi.org/10.1007/978-3-642-17819-1_34

Groth P, Gil Y, Cheney J, Miles S (2012) Requirements for provenance on the Web. Int J Digit Curat 7(1):39–56. https://doi.org/10.2218/ijdc.v7i1.213

Haber S, Stornetta WS (1991) How to time-stamp a digital document. J Cryptol (3):99–111. https://doi.org/10.1007/BF00196791

Haeberlen A, Kouznetsov P, Druschel P (2007) PeerReview: practical accountability for distributed systems. ACM SIGOPS Oper Syst Rev 41(6):175–188. https://doi.org/10.1145/1294261.1294279

Halpern JY, Pearl J (2005a) Causes and explanations: a structural-model approach. Part I: Causes. Br J Philos Sci 56(4):843–887

Halpern JY, Pearl J (2005b) Causes and explanations: a structural-model approach. Part II: Explanations. Br J Philos Sci 56(4):889–911

Heussner KM (2009) Woman loses benefits after posting Facebook pics. ABC News 23. https://abcnews.go.com/Technology/AheadoftheCurve/woman-loses-insurance-benefits-facebook-pics/story?id=9154741

Ilievski F, Garijo D, Chalupsky H, Divvala NT, Yao Y, Rogers C, Li R, Liu J, Singh A, Schwabe D, et al. (2020) KGTK: a toolkit for large knowledge graph manipulation and analysis. https://arxiv.org/abs/2006.00088

Jain P, Jain P, Kumaraguru P (2013) Call me maybe: understanding nature and risks of sharing mobile numbers on online social networks. In: Proceedings of the first ACM conference on online social networks, pp 101–106. https://doi.org/10.1145/2512938.2512959

JEITA (2002) Exchangeable image file format for digital still cameras: Exif version 2.2

Jernigan C, Mistree BFT (2009) View of gaydar: Facebook friendships expose sexual orientation. https://firstmonday.org/article/view/2611/2302

Jones HC (2007) XHTML documents with inline, policy-aware provenance. PhD thesis, Massachusetts Institute of Technology. https://dspace.mit.edu/handle/1721.1/41627

Jurgens D, Tsvetkov Y, Jurafsky D (2017) Writer profiling without the writers text. In: Ciampaglia GL, Mashhadi A, Yasseri T (eds) Social informatics, Part II. Springer, Cham, pp 537–558. https://doi.org/10.1007/978-3-319-67256-4_43

Keküllüoglu D, Magdy W, Vaniea K (2020) analyzing privacy leakage of life events on Twitter. In: 12th ACM conference on web science, pp 287–294. https://doi.org/10.1145/3394231.3397919

Keküllüoğlu D, Magdy W, Vaniea K (2020) analyzing privacy leakage of life events on Twitter. In: 12th ACM conference on web science, pp 287–294. https://doi.org/10.1145/3394231.3397919

Kumari P, Pretschner A, Peschla J, Kuhn JM (2011) Distributed data usage control for web applications: a social network implementation. In: Proceedings of the first ACM conference on data and application security and privacy, pp 85–96. https://doi.org/10.1145/1943513.1943526

Küsters R, Truderung T, Vogt A (2010) Accountability: definition and relationship to verifiability. In: Proceedings of the 17th ACM conference on computer and communications security, pp 526–535. https://doi.org/10.1145/1866307.1866366

Lessig L (2003) The creative commons. Fla L Rev 55:763

Longo DL, Drazen JM (2016) Data sharing. N Engl J Med 374(3):276–277. https://doi.org/10.1056/NEJMe1516564

Magliacane S (2012) Reconstructing provenance. In: Cudré-Mauroux P, Heflin J, Sirin E, Tudorache T, Euzenat J, Hauswirth M, Parreira JX, Hendler J, Schreiber G, Bernstein A, Blomqvist E (eds) The semantic web – ISWC 2012. Springer, Heidelberg, pp 399–406. https://doi.org/10.1007/978-3-642-35173-0_29

Manovich L (2005) Remixing and remixability. http://www.manovich.net/DOCS/Remix_modular.doc

Marshall MS, Boyce R, Deus HF, Zhao J, Willighagen EL, Samwald M, Pichler E, Hajagos J, Prudhommeaux E, Stephens S (2012) Emerging practices for mapping and linking life sciences data using RDF—a case series. J Web Semant 14:2–13. https://doi.org/10.1016/j.websem.2012.02.003

McGuinness DL, Da Silva PP (2004) Explaining answers from the Semantic Web: The Inference Web approach. J Web Semant 1(4):397–413. https://doi.org/10.1016/j.websem.2004.06.002

Moreau L, Groth P, Miles S, Vazquez-Salceda J, Ibbotson J, Jiang S, Munroe S, Rana O, Schreiber A, Tan V, et al. (2008) The provenance of electronic data. Commun ACM 51(4):52–58. https://doi.org/10.1145/1330311.1330323

Oransky I, Marcus A (2016) Criticism of 'research parasites' moves NEJM in the wrong direction. https://www.statnews.com/2016/01/26/research-parasites-nejm/

Pretschner A, Hilty M, Basin D (2006) Distributed usage control. Commun ACM 49(9):39–44. https://doi.org/10.1145/1151030.1151053

Saltzer JH, Schroeder MD (1975) The protection of information in computer systems. Proc IEEE 63(9):1278–1308. https://doi.org/10.1109/PROC.1975.9939

Samavi R, Consens MP (2012) L2TAP+ SCIP: an audit-based privacy framework leveraging Linked Data. In: 8th international conference on collaborative computing: networking, applications and worksharing, IEEE, pp 719–726. https://doi.org/10.4108/icst.collaboratecom.2012.250607

Seneviratne OW (2012) Augmenting the Web with accountability. In: Proceedings of the 21st international conference on World Wide Web, pp 185–190. https://doi.org/10.1145/2187980.2188006

Seneviratne OW (2014) Accountable systems: enabling appropriate use of information on the Web. PhD thesis, Massachusetts Institute of Technology. https://dspace.mit.edu/handle/1721.1/93833

Senevitane O, Kagal L (2011) Addressing data reuse issues at the protocol level. In: 2011 IEEE international symposium on policies for distributed systems and networks, IEEE, pp 141–144. https://doi.org/10.1109/POLICY.2011.48

Seneviratne O, Kagal L (2014) Enabling privacy through transparency. In: Twelfth annual international conference on privacy, security and trust, IEEE, pp 121–128. https://doi.org/10.1109/PST.2014.6890931

Seneviratne O, Monroy-Hernandez A (2010) Remix culture on the Web: a survey of content reuse on different user-generated content websites. In: ACM Conference on Web Science

Seneviratne O, Kagal L, Berners-Lee T (2009) Policy-aware content reuse on the Web. In: Bernstein A, Karger DR, Heath T, Feigenbaum L, Maynard D, Motta E, Thirunarayan K (eds) The semantic web – ISWC 2009. Springer, Heidelberg, pp 553–568. https://doi.org/10.1007/978-3-642-04930-9_35

Simmhan YL, Plale B, Gannon D (2005) A survey of data provenance in e-Science. ACM Sigmod Rec 34(3):31–36. https://doi.org/10.1145/1084805.1084812

Voigt P, von dem Bussche A (2015) The EU general data protection regulation (GDPR): a practical guide. Springer, Cham. https://doi.org/10.1007/978-3-319-57959-7

Wash R (2010) Folk models of home computer security. In: Proceedings of the sixth symposium on usable privacy and security. https://doi.org/10.1145/1837110.1837125

Yang D, Kraut RE, Smith T, Mayfield E, Jurafsky D (2019) Seekers, providers, welcomers, and storytellers: Modeling social roles in online health communities. In: Proceedings of the 2019 CHI conference on human factors in computing systems. https://doi.org/10.1145/3290605.3300574

Zamal FA, Liu W, Ruths D (2012) Homophily and latent attribute inference: inferring latent attributes of Twitter users from neighbors. In: Proceedings of the sixth international AAAI conference on weblogs and social media, AAAI, pp 387–390. https://www.aaai.org/ocs/index.php/ICWSM/ICWSM12/paper/viewFile/4713/5013

Chapter 3
The Right (Provenance) Hammer for the Job: A Comparison of Data Provenance Instrumentation

Adriane Chapman ⓘ, Abhirami Sasikant, Giulia Simonelli, Paolo Missier ⓘ, and Riccardo Torlone ⓘ

3.1 Introduction

The W3C Provenance Working Group defines provenance as the "information about entities, activities and people involved in producing a piece of data or thing, which can be used to form assessments about its quality, reliability or trustworthiness" (Mor 2013b,a). This statement by design gives no indication of how much information is needed to perform these assessments. Past examples have included as little as an original source (see ISO 19115-1:2014)[1] and as much as a full chain of processing for an individual data item (Brauer et al. 2014). While the collection and processing of provenance is important: to assess quality (Huynh et al. 2018), enable reproducibility (Thavasimani et al. 2019), reinforce trust in the end product (Batlajery et al. 2018), or to aid in problem diagnosis and process debugging (Herschel et al. 2017), what provenance is enough, and is it worth the cost of instrumentation?

---

[1]https://www.iso.org/obp/ui/#iso:std:iso:19115:-1:ed-1:v1:en

A. Chapman (✉) · A. Sasikant
University of Southampton, Southampton, UK
e-mail: adriane.chapman@soton.ac.uk; as1n16@soton.ac.uk

G. Simonelli · R. Torlone
Università Roma Tre, Rome, Italy
e-mail: giullia.simonelli@uniroma3.it; riccardo.torlone@uniroma3.it

P. Missier
Newcastle University, Newcastle upon Tyne, UK
e-mail: paolo.missier@ncl.ac.uk

© Springer Nature Switzerland AG 2021
L. F. Sikos et al. (eds.), *Provenance in Data Science*, Advanced Information and Knowledge Processing, https://doi.org/10.1007/978-3-030-67681-0_3

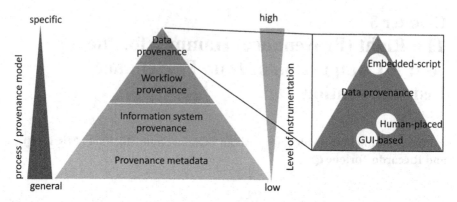

Fig. 3.1 Reproduction of Provenance Hierarchy from Herschel et al. (2017) and expanded to highlight the methods to instrument data provenance in the Orange3 framework in this work

Initial work and definitions of provenance (Buneman et al. 2001) evolved into many different types of provenance, including how, why, where (Cheney et al. 2009), and why not (Chapman and Jagadish 2009). Many surveys exist that create categorizations of different types of provenance, including provenance found in e-Science platforms (Simmhan et al. 2005), for computational processing (Freire et al. 2008), data provenance (Glavic and Dittrich 2007), and scripts (Pimentel et al. 2019). Based on a recent survey focused on looking at provenance and instrumentation needs (Herschel et al. 2017), provenance can be broadly classified into the following types: metadata, information system, workflow, and data provenance. In this context, Herschel et al. identified that with more specific information, such as data provenance, the level of instrumentation increases (Herschel et al. 2017).

Even within the "data provenance" category, there is a huge range in terms of instrumentation options available which can change the granularity of the provenance available for use. Within the hierarchy presented in Herschel et al. (2017), we investigate differences in instrumentation and supported queries of data provenance (Fig. 3.1).

In this work, we focus on a very specific set of tasks: collecting provenance of machine learning pipelines. There is a large amount of work involved in gathering and preparing data for use within a machine learning pipeline. Which data transformations are chosen can have a large impact on the resulting model (Feldman et al. 2015; Zelaya et al. 2019; Zelaya 2019). Provenance can help a user debug the pipeline or reason about final model results. In order to investigate granularity of data provenance and instrumentation costs, we use the application *Orange*[2] (Demšar et al. 2013) and show how provenance instrumentation choices can greatly affect the types of queries that can be posed over the data provenance captured. Our contributions include:

---

[2]https://orange.biolab.si

- We identified a tool, Orange3, that assists in organizing Python scripts into machine learning workflows, and use it to:
  - Identify a set of use cases that require provenance;
  - Constrain the environment so that a set of provenance instrumentation techniques can be compared.
- Implemented provenance instrumentation using a GUI-based insertion, embedded within scripts, and via expert hand-encoding in a machine learning pipeline.
- Compared the use of the resulting data provenance for each of these instrumentation approaches, in order to address issues the real-world use cases found in Orange3.

We begin in Sect. 3.2 with an overview of the Orange3 tool, and the provenance needs expressed by end users. In Sect. 3.3, we provide a brief overview of instrumentation options; Sects. 3.3.1, 3.3.2, and 3.3.3 discuss an instrumentation of ML pipeline scripts by instrumenting either hand-coded scripts (Sect. 3.3.1); the Orange3 GUI (Sect. 3.3.2); or by embedding directly in the scripts (Sect. 3.3.3). In Sect. 3.4, we compare these approaches with respect to solving the set of real-world problems identified in Sect. 3.2. We conclude in Sect. 3.5.

3.2   Case Study: Machine Learning Pipelines and Orange3

Machine learning is more than just the algorithms. It encompasses all of the data discovery, cleaning, and wrangling required to prepare a dataset for modeling and relies upon a person constructing a pipeline of transformations to ready the data for use in a model (Shang et al. 2019). In 1997, the development of Orange began at the University of Ljubljana. The goal was to address difficulties in illustrating aspects of machine learning pipelines with a standalone command-line utility. Since then, Orange has been in continuous development (Demšar et al. 2013), with the most recent release being Orange3.

Orange is an open-source data visualization, machine learning, and data mining toolkit. It provides an interactive visual programming interface for creating data analysis and machine learning workflows. It helps modelers gain insights from complex data sources. Orange can be utilized as a Python library, where Python scripts can be executed on the command line. The Orange package enables users to perform data manipulation, widget alterations, and add-on modeling. This platform caters to users of different levels and backgrounds, allowing machine learning to become a toolkit for any developer, as opposed to a specialized skill.

Orange features a canvas, a visual programming environment shown in Fig. 3.2, in which users create machine learning workflows by adding pipeline components called widgets. Widgets provide basic self-contained functionalities for all key stages of a machine learning pipeline. Each of these widgets is classified into categories based on their functionality and assigned priority, and is listed in the Orange toolbox. In addition, each widget has a description, and input/output

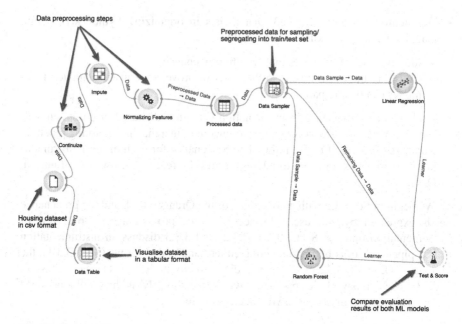

Fig. 3.2 Orange3 Canvas to support construction of machine learning pipelines. The pipeline shown is the "Housing" pipeline whose provenance is later shown in Figs. 3.3 and 3.7

channels associated with it. These widgets communicate and pass data objects, by means of communication channels. A data analysis workflow in Orange can be defined as a collection of widgets and communication channels.

The Orange library is a "hierarchically-organized toolbox of data mining components" (Demšar et al. 2013). The lower-level procedures are placed at the bottom of the hierarchy such as data processing and feature scoring, and higher-level procedures such as classification algorithms are placed at the top of hierarchy. The main branches of component hierarchy include data management and preprocessing, classification, regression, association, ensembles, clustering, evaluation, and projections. This library simplifies the creation of workflows and the generation of data mining approaches by enabling combinations of existing components. The key strength of Orange lies in connecting widgets in numerous ways by adopting various methodologies, resulting in new schemata (Demšar et al. 2013).

Ultimately, Orange allows users to organize and execute scripts that when placed together create machine learning pipelines.

3.2.1 Provenance Needs in Orange3

In general, provenance can be used to assess quality (Huynh et al. 2018), reproduce scientific endeavors (Thavasimani et al. 2019), facilitate trust (Batlajery et al. 2018), and help with debugging (Herschel et al. 2017). In this work, we focus on the debugging aspect. In order to gather real-world use cases of provenance in this domain, we reviewed a total of 370 use cases from the following forums: Data Science Stack Exchange (DSSE), Stack Overflow (SO), and the FAQ section on the Orange website (FAQ). From these 370 use cases, 12 use cases were considered relevant to the workflow debugging processes, and are listed in Table 3.1.

3.3   Overview of Instrumentation Possibilities

The Orange tool was created to facilitate creation of machine learning pipelines. While not implemented in Orange, there are some previous works on provenance in machine learning pipelines. Vamsa captures provenance of the APIs called and libraries used within a particular ML pipeline in order to help the user debug the pipeline (Namaki et al. 2020). However, the provenance captured does not focus on the data and what happened to it, instead on how the pipeline is constructed and the organization of scripts within it. Smaller than a pipeline, Lamp gathers provenance using a graph-based machine learning algorithm (GML) to reason over the importance of data items within the model (Ma et al. 2017). Finally, Jentzsch and N. Hochgeschwender recommend using provenance when designing ML models to improve transparency and explainability for end users (Jentzsch and Hochgeschwender 2019).

Because this work focuses on data provenance, we do not specifically review instrumentation options for script management (Pimentel et al. 2019), social interactions (Packer et al. 2019), or provenance from relational systems (Green et al. 2007, 2010). Provenance of scripts has recently been surveyed (Pimentel et al. 2019), with a classification of annotations, definition, deployment, and execution. In this work, the focus is effectively on *execution provenance*, "the origin of data and its derivation process during execution" of scripts.

Scientific workflows have been the earliest adopters of provenance in scripts, and we look to this community for inspiration in a machine learning pipeline creation setting. According to Sarikhani and Wendelborn (2018), provenance collection mechanisms have the ability to access distinct types of information in scientific workflow systems at the workflow level, activity level, and operating system level.

- Workflow-level provenance is captured at the level of scientific workflow systems. Here, the provenance capture mechanism is either attached to or is integrated within the scientific workflow system itself. A key advantage of this approach is that, the mechanism is closely coupled with the workflow system, and thus enables a direct capture process through the systems API. Within our machine learning pipeline creation context, this would be similar to capturing the design and configuration of the pipeline. We investigate this in Sect. 3.3.2.
- Activity(process)-level provenance captures provenance at the level of processes or activities of the scientific workflow. Here, the provenance mechanism is

Table 3.1 Uses for provenance identified in Stack Overflow (SO), Data Science Stack Exchange (DSSE)

QID	Source	Type	Description	Provenance needed
UC1	DSSE	Prediction	When applying the "Predictions" widget on the same training dataset, the results (i.e. probability scores) are different	The set of processes and their ordering in the execution.
UC2	DSSE	Prediction	When applying the "Impute" widget during preprocessing on train/test dataset, same values are predicted for all rows	The set of processes and their ordering in the execution.
UC3	DSSE	Classification/prediction	After performing image classification using an ML model, prediction probabilities are constant on test images	The processes and data of the execution; actual data changes required.
UC4	DSSE	Verification	From a constructed workflow using image classification (add-on widgets), ascertain whether the workflow performs "transfer learning"	The processes and data of the execution; actual data changes required.
UC5	SO	Prediction/Evaluation	Data fed to a handful of classifiers for comparing evaluation results fails to update, when using the "Rank" and "Test Learner" widgets in the workflow	The processes and data of the execution; the structure of the workflow.
UC6	DSSE	Rectification	Disproportionate allocation of labels after performing data analysis and modelling (inaccurate classification accuracy)	Fine-grained provenance showing detailed pipeline working to aid in logical understanding of widget utilization; actual data changes required.
UC7	DSSE	Prediction	Inaccuracy in the prediction of target variable using k-NN and linear regression ML models in an Orange workflow	The processes and data of the execution; actual data changes required
UC8	DSSE	Pipeline Construction	Ambiguity/difficulty in loading a data file (setting up the corpus) onto the Orange platform to perform data modelling using a linear regression model	Collective provenance of multiple workflows using the same ML model that would aid in predicting/suggesting possible widget combinations (paths)

UC9	DSSE	Prediction/Evaluation	Application of the "Test and Score" and "Predictions" widget on the same data utilizing the same ML model; produces differing results	The processes and data of the execution; actual data changes required
UC10	DSSE	Prediction	Differences in the predictions and corresponding goodness-of-fit R2 metric for the linear regression model on Orange and scikit-learn	The processes and data of the execution; actual data changes required
UC11	SO	Reproducibility of results	Differing results, clusters, and corresponding silhouette scores for the same processed dataset (including preprocessing steps); when using the "k-means" ML model widget	Widget connections and processes of the workflow; actual data changes are also required
UC12	SO	Evaluation	Deviations in the variance percentages accounted for by each of the principal components of the same dataset; for Principal Component Analysis (PCA) on Orange and scikit-learn	The processes and data of the execution; actual data changes required

independent from the scientific workflow system. In this approach, the mechanism requires relevant documentation, relating to information derived from autonomous processes, for each step of the workflow. In our context, this would be similar to capturing the information about the exact script that ran and what occurred to the data during that execution. This will be discussed in Sect. 3.3.3.

- Operating system-level provenance is captured in the operating system. This approach relies on the availability of a specific functionality at the OS level, and thus requires no modifications to existing scripts or applications. In comparison to workflow level, this mechanism captures provenance at a finer granularity. We omit this level here because the provenance at this level is unlikely to answer immediate user queries.

Looking beyond workflow systems, we highlight some of the classic architectural options (Allen et al. 2010) that could be used within the machine learning context and Orange.

Log Scraping

Log files contain much information that can be utilized as provenance. Collecting provenance information from these log files involves a log file parsing program. An example of this technique can be seen in Roper et al. (2020) in which the log files (change sets and version history files) are used to construct provenance information. Using log files means that developers keen on instrumenting provenance capture do not need to work within possibly proprietary or closed systems. On the other hand, there is little control as to the type and depth of information that can be obtained from log information. This approach could be used to gather either workflow-level or activity-level provenance from Sarikhani and Wendelborn (2018) depending on the set of logs available.

Human Supplied

The humans who are performing the work can sometimes be tasked to provide provenance information. Users of RTracker (Lerner et al. 2018) demonstrated that they were invested enough to carefully define and model the provenance needed in their domain. YesWorkflow (McPhillips et al. 2015; Zhang et al. 2017) allows users to embed notations within the comments of a code that can be interpreted by YesWorkflow to generate the provenance of scripts. These approaches effectively capture workflow-level provenance from Sarikhani and Wendelborn (2018). See Sect. 3.3.1 for human-supplied provenance in the Orange framework.

Application Instrumentation

A straightforward method for instrumenting provenance capture in any application is to modify the application directly. The major benefit of this approach is that the developer can be very precise regarding the information captured. The drawback is that the application must be open for developers to modify, and all subsequent development and maintenance must also take provenance into account. Applications that are provenance-aware cover the range from single applications, such as provenance for visualization builder (Psallidas and Wu 2018), to larger workflow systems (Koop et al. 2008; Missier and Goble 2011; Santos et al. 2009; Souza et al. 2017). Other examples include provenance capture within MapReduce (Ikeda et al. 2012), Pig Latin (Amsterdamer et al. 2011), or Spark (Guedes et al. 2018; Interlandi et al. 2015; Psallidas and Wu 2018; Tang et al. 2019). This approach could be used to gather either workflow-level or activity-level provenance (Sarikhani and Wendelborn 2018). See Sect. 3.3.2 for application-based provenance capture in the Orange framework.

Script Embedding
Several approaches exist to embed directly into scripts. For example, NoWorkflow (Murta et al. 2015; Pimentel et al. 2017, 2016a) embeds directly into Python scripts and automatically logs provenance by program slicing. Other approaches use function-level information in Java to capture provenance information (Frew et al. 2008). In Sarikhani and Wendelborn (2018), this would be most appropriate for activity-level provenance capture. See Sect. 3.3.3 for script-embedding provenance capture in the Orange framework.

Recent work looks at different instrumentation vs. provenance-content choices available in systems that help scientists collect provenance of scripts (Pimentel et al. 2016a). NoWorkflow (Pimentel et al. 2017, 2016b) captures script execution by program slicing, while YesWorkflow (McPhillips et al. 2015; Zhang et al. 2017) asks users to hand-create capture at the desired places. The two systems are brought together in Pimentel et al. (2016a), and the types of provenance information have been compared. In the following sections, we look more closely at human supplied, application instrumentation and script embedding to gather provenance within the Orange toolkit.

3.3.1 Human-Supplied Capture

In order to understand what a human with basic provenance instrumentation tools at their disposal would do, we asked an expert provenance modeler to create a machine learning pipeline to predict median house values using linear regression and random forest on a housing dataset, and to instrument the code for provenance. A diagrammatic view of this workflow is shown in the Orange3 canvas in Fig. 3.2. This in effect reflects the baseline of what is currently obtainable by an invested human who has provenance modeling skills and is willing to take the time to insert calls to a provenance capture API, with carefully chosen information about processes, data, and agents. In this work, the expert used the standard *Prov Python*[3] libraries. Figure 3.3 shows the provenance generated by the expert for this pipeline.

---

[3] https://prov.readthedocs.io

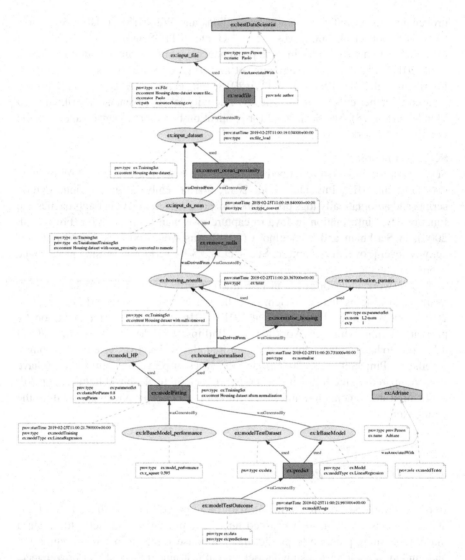

Fig. 3.3 Example of provenance captured via an expert provenance modeler for the machine learning pipeline

3.3.2 GUI-Based Capture

The provenance capture mechanism described in this section is built for Orange 3.16, released in October 2018. In order to capture provenance of the set of operations performed in Orange, while minimizing developer input in the creation and maintenance of provenance capture, we utilize the inherent class structure of the widgets used in the GUI.

Table 3.2 A subset of operations offered by Orange3 to build machine learning pipelines. The hook used by both the GUI and embedded approaches is noted

Orange3		GUI	Embedded
Type	Operator	Widget class	Transformation type
Data	File	OWFile	–
	SQL Table	OWSql	
	Select columns	OWSelectAttributes	Dimensionality reduction
	Select rows	OWSelectRows	
	Select relevant features	OWPreprocess	
	Select random features	OWPreprocess	
	Select data by index	OWSelectByDataIndex	
	Purge domain	OWPurgeDomain	
	Discretize	OWDiscretize	Feature transformation
	Continuize	OWContinuize	
	Randomize	OWRandomize	
	Impute	OWImpute	Imputation
	Edit domain	OWEditDomain	Value transformation
	Feature constructor	OWFeatureConstructor	Space transformation
	Create class	OWCreateClass	
Model	SVM		–
	Linear regression		
	kNN		
	Tree	OWBaseLearner	
	Stochastic gradient descent		
	Random forest		
	22 Other models		
Visualize	Box plot	OWBoxPlot	–
	Scatter plot	OWScatterPlot	

OWWidget, the parent widget class, is extended by all widget classes. This class provides all basic functionality of widgets regarding inputs, outputs, and methods that are fundamental to widget functioning. After analyzing the code associated with different widget categories, it was observed that this idea could be applied only to groups of widgets, with similar functionality such as model and visualization widgets; the parent classes extend OWWidget class for such widget groups. For example, as shown in Table 3.3, OWBaseLearner extends OWWidget and is the parent to all model widgets. Instrumenting OWBaseLearner for provenance capture provides provenance functionality to all modeling widgets.

For other categories of widgets, it is necessary to capture provenance in each respective widget class. Because each of these widgets contains different input/output signal types and contents, the capture of this information is not standardized. Table 3.2 contains the set of Widget classes that are utilized to gather information for each operation. A diagrammatic representation of provenance cap-

Table 3.3 Parameters and provenance captured for each function type. (A) activities recorded; (E) entities recorded; (R) relationships recorded

Function type	Parameters	Provenance contains
Dimensionality reduction	out_dataframe description	A: a single activity, f, is created.
		R: the original entities are *invalidatedBy* f.
Feature transformation	out_dataframe columnsName description	A: a single activity, f, is created.
		E: a new entity is created for each modified item.
		R: the new entity *wasGeneratedBy* f; the new entity *wasDerivedFrom* the original entity; f *used* the set of feature items; the original entity *wasInvalidatedBy* f.
Space transformation	out_dataframe columnsName description	A: a single activity, f, is created.
		E: a new entity is created for every new attribute.
		R: f *uses* the set of entities that belong to the features used for the transformation; the new entities are *generatedBy* f and *derivedFrom* the entities of the related record and the features used for the transformation.
Instance generation	out_dataframe description	A: a single activity, f, is created.
		E: a new entity is created for every new instance.
		R: each new entity *wasGeneratedBy* f; f *uses* the existing entities that belong to the related feature.
Imputation (Dependent)	out_dataframe isIndependent=F description	A: a new activity for each feature is created.
		E: a new entity is created for every replaced value.
		R: each activity *uses* the entity for the related non-null feature; null entities are *invalidated*.
Imputation (Independent)	out_dataframe isIndependent=T description	A: a new activity for each feature is created.
		E: a new entity is created for every replaced value.
		R: each activity *uses* the entity for the related non-null feature; null entities are *invalidated*.
Value transformation	out_dataframe value description	A: a single activity, f, is created.
		E: a new entity is created for every replaced value.
		R: f *uses* the set of entities that belong to the features used for the transformation; the new entities are *generatedBy* f and *derivedFrom* the entities of the related record and the features used for the transformation.

ture involving OWWidget base class and code example of widget-based provenance capture are shown in Fig. 3.4. For more implementation details, please refer to Sasikant (2019).

In this section, we identify a capture point within the Orange3 architecture that allows provenance to be automatically captured as code instantiated by widgets gets executed. While this instrumentation is a "light touch" and can weather additional widgets that belong to predefined classes, it has no real insight as to what is happening to the data itself. By capturing at the GUI level, as the user drags and

a   **b**

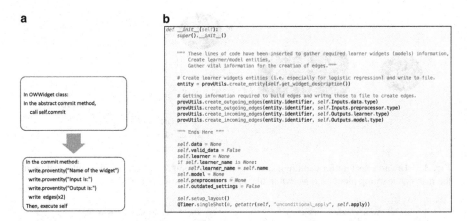

Fig. 3.4 Provenance capture instrumentation via the Orange3 GUI. (**a**) Diagrammatic overview of GUI-based capture. (**b**) Example code required to add provenance capture to Orange widgets

drops operators, the provenance generated can be both retrospective and prospective (Lim et al. 2010).

3.3.3   Embedded-Script Capture

While the approach identified in Sect. 3.3.2 is a "light touch," it is not resilient with respect to Orange3 code changes and additions. It also does not provide the ability to introspect on changes to the data itself. In this approach, we instrument the Python scrips that execute the data preprocessing called upon execution of the pipeline.

Abstracting the functionality of the scripts that are executed by the Orange3 framework and represented by the Widgets in Orange3, there are several categories of functionality based on how the data itself is impacted or changed. These include dimensionality reduction, feature transformation, space transformation, instance generation, imputation, and value transformation, as shown in Table 3.2. Because script-embedded instrumentation is required to capture what happens with the data, and each type of operation does a different type of transform over the data, there are different provenance instrumentation calls for each type. However, the same type of instrumentation can then be reused for other scripts of the same category as shown in Table 3.3. Unfortunately, despite the abstract reuse of type of capture, every script must be individually provenance-instrumented. The architecture is shown in Fig. 3.5, and details of this implementation can be found in Simonelli (2019). Figure 3.6 contains a sample provenance record for a value transformation operation.

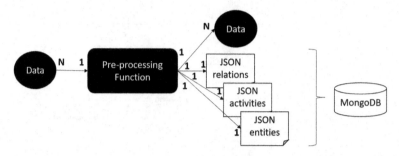

Fig. 3.5 The architecture for capture instrumentation for fine-grained provenance. Information in the machine learning pipeline is shown in black. Provenance artifacts are white

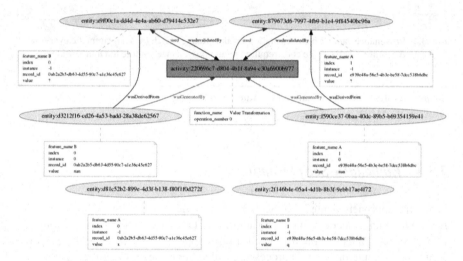

Fig. 3.6 Example of provenance captured for just one operation, value transformation, with a deeper instrumentation

3.4 Comparison of Instrumentation Approaches

We compare the data provenance and instrumentation requirements of our three approaches looking at provenance content (Sect. 3.4.1), the ability to answer real-world questions (Sect. 3.4.2), and the pros and cons of the instrumentation approach (Sect. 3.4.3).

3.4.1 Provenance Collected

We begin by comparing the three approaches with respect to the content of the provenance. The same machine learning pipeline that the expert provenance enabled

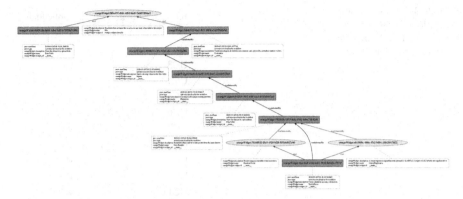

Fig. 3.7 Example of provenance captured via the GUI-based capture for the same machine learning pipeline that produced Fig. 3.3

was instantiated in Orange3, as shown in Fig. 3.2. The GUI-based provenance capture was applied, and produced the provenance shown in Fig. 3.7. On comparing the expert provided provenance graph (EP) in Fig. 3.3 with the GUI-generated provenance graph (GP) shown in Fig. 3.7 and the Script generated provenance (SP), the following can be seen:

- **Agents**: EP captured the presence of two agents. GP and SP do not have agents, as the Orange toolkit only involves one user at a time, who would be responsible for constructing the workflow.
- **Granularity of processes**: EP captured provenance at fine levels of granularity, providing minute details regarding processes involved and transformations of the specific dataset. For instance, the process of reading data from a file has been separated into three sub-stages: Entity [Input file] → Activity [Read file] → Entity [Input dataset]. Hence, provenance has been intricately captured in this scenario. On the other hand, GP and SP capture provenance at coarser granularity, providing key information of processes at a higher level. For the same process of reading data from file, provenance is captured widgetwise in GP, as widgets are building blocks of Orange workflows: Entity [File widget] → Activity [Data Table widget]. Hence, provenance of communication between widgets provides vital information (input/output signal components, widget details). However, GP does not capture provenance regarding data content changes, due to inaccessibility to the controller of Orange, and since provenance is captured at an abstract level, while SP captures large amounts of provenance about specific data changes.
- **Entity vs. activity**: Modeling decisions for entities and activities are different in EP, GP, and SP because of the organization imposed by the environment in which the capture calls were implemented.

In all approaches, provenance of all fundamental stages of the machine learning pipeline are captured: loading dataset, preprocessing/cleaning data, training models,

Table 3.4 Comparison of abilities to answer queries based on the information captured by a particular technique

ID	Human	GUI	Embedded
UC1	✓	✓	✓
UC2	✓	✓	✓
UC3	–	–	✓
UC4	–	–	✓
UC5	✓	✓	✓
UC6	–	–	✓
UC7	–	–	✓
UC8	✓	✓	✓
UC9	–	–	✓
UC10	–	–	✓
UC11	–	–	✓
UC12	–	–	✓

testing models on test data, and analysis of model predictions. There are some differences relating to detail in the provenance model, but for the most part, the three approaches capture the same type of information, with variations in detail and modeling.

3.4.2   Answering Provenance Queries

Returning to the Real-World use of provenance for machine learning pipeline development and debugging as described in Sect. 3.2.1, we now review which of these use cases can actually be resolved by the provenance captured via the three methods discussed in this work. Table 3.4 shows the Orange3 user queries that can be answered with each technique.

In essence, UC1, UC2, UC5, and UC8 contain questions that can be answered by understanding the overall pipeline design. They can be answered by understanding which data preprocessing functions were utilized, and the order in which preprocessing steps were made. The remainder of the use cases requires an analysis over the data itself, particularly spread and changes of spread in the data based on preprocessing.

3.4.3   The Cost of Provenance Instrumentation

Table 3.5 provides a high-level overview of the discussion within this section. There are two distinct roles that should be considered when contemplating data provenance instrumentation and usage: provenance modeler and provenance user. While in many cases, these are the same individual, the two roles require very different skills. A provenance modeler must:

Table 3.5 Comparison of costs for each method

	Human-placed	GUI-based	Embedded script
Developer/end user	Same	Can be different	Can be different
Num. files processed	Num. scripts written	≤ Num. operators	Num. operators
Fragility of instrumentation	↓	↑	↓
Requires tool openness	↓	↑	↓
Constrains user to tool	↓	↑	↓
Size/detail of provenance	↓	↓	↑

- Understand how to model provenance. This includes the following:
  - Understand provenance, why it is used, and the important information required for capturing provenance including objects and relationships.
  - Understand the standards that are available.
  - Understand the application to be provenance enabled, and how the concepts in that application relate to provenance objects and relationships.
- Write the code to instrument the application.
- Maintain the provenance instrumentation throughout application updates.

  The provenance user must:

- Interpret the provenance that is returned by the system to understand how it answers a given problem.

In many cases, particularly scientific exploration systems, these two roles are held by the same person. However, in large systems obtained through a formal acquisition process, e.g., for governments and large organizations, these two roles are filled by different people who may never interact with each other. In the context of machine learning pipelines and provenance instrumentation discussed in this work, human-supplied provenance requires that the roles of provenance modeler and provenance user are indeed held by the same person. Both the GUI-based and the Script-embedded allow the roles to be held by different individuals. This allows the end user of the provenance to be essentially provenance-unaware if they choose to be: they can merely be a consumer of data and not a developer.

However, there is a difference between GUI-based and script-based implementations. The GUI-based implementations, while providing less information about the data to the end user, are abstract enough that the provenance modeler comes up with fewer insertion points for capture calls within the 3rd-party code. In the worst case, GUI-based will insert as many capture calls as the number of operations, just as in the script-based implementations.

However, because we can utilize good code design, and occasionally embed the call in a parent class, it is sometimes possible for the GUIbased to have fewer calls than the script-based. This was done for OWBaseLearner, in which 1 file must be provenance enabled in the GUI-based method in order to capture provenance in 28 different learning models (see Table 3.2). Unless there is an upgrade to the GUI itself, any number of scripts and processing functions can be added to the underlying Orange3 tool, and the provenance modeler does not need to be aware of them. The GUI-based design is more fragile with respect to Orange3 refactoring and code updates. Unlike the script-embedding approach in which the fundamental machine learning Python scripts are provenance enabled, and suffer very little churn, the front end has experienced many code refreshes. Indeed, during the writing of this, the Orange Framework was undergoing yet another refactoring that likely changes the GUI-based provenance capture completely.

More generally and not constrained within Orange, the three approaches differ based on whether a tool is open or not. A human can embed provenance capture within their own scripts, and most underlying scripts are open. However, aiming for an application capture, like the Orange3, GUI-based approach requires that tool to be open. In the case of Orange3, this is true, but it may not be for many other applications. In a similar analysis, both human-based and script embedding allow a user to choose their tools of choice, while application-embedded implementations require the user to work within that single tool. Finally, the size of the provenance is similar between human-generated (e.g., Fig. 3.3 and GUI-based Fig. 3.7) implementations. Where the human is limited by time and inclination, the application is often limited by available hooks to information. In contrast, script-embedded capture can see and record all of the details, resulting in a larger provenance record (e.g., a small subset in Fig. 3.5).

3.5 Conclusions

In this work, we focused on an analysis of the types of data provenance that can be captured via very different instrumentation approaches within the same "task." In such a task, a user builds a machine learning pipeline and attempts to debug it. We identified many real-world scenarios of this process, documented in software development forums. We restricted ourselves to a set of Python scripts for processing data for machine learning, and a GUI-based tool, Orange, for organizing them. We showed how choices of provenance capture can greatly affect the types of queries that can be posed over the data provenance captured. We implemented provenance instrumentation using a GUI-based insertion, embedded within Python scripts, and via careful hand-encoding, applied to building a machine learning pipeline. We highlighted the instrumentation approaches possible by analyzing those used in scientific workflows, and described the different implementation approaches used within our narrowed domain. We then compared the utility of the resulting data provenance for each of these instrumentation approaches, to answer the real-world use cases found in Orange3. The results of this work provide a comparative analysis for future developers to identify and choose an apt instrumentation approach for future efforts.

Acknowledgments This work was partially supported by EPSRC (EP/S028366/1).

References

Allen MD, Seligman L, Blaustein B, Chapman A (2010) Provenance capture and use: a practical guide. the MITRE Corporation. https://www.mitre.org/sites/default/files/publications/practical-provenance-guide-MP100128.pdf

Amsterdamer Y, Davidson SB, Deutch D, Milo T, Stoyanovich J, Tannen V (2011) Putting lipstick on Pig: enabling database-style workflow provenance. In: Proceedings of the VLDB endowment, pp 346–357. https://doi.org/10.14778/2095686.2095693

Batlajery BV, Weal M, Chapman A, Moreau L (2018) Belief propagation through provenance graphs. In: Belhajjame K, Gehani A, Alper P (eds) Provenance and annotation of data and processes. Springer, Cham, pp 145–157. https://doi.org/10.1007/978-3-319-98379-0_11

Brauer PC, Czerniak A, Hasselbring W (2014) Start smart and finish wise: the Kiel Marine Science provenance-aware data management approach. In: 6th USENIX Workshop on the Theory and Practice of Provenance. https://www.usenix.org/system/files/conference/tapp2014/tapp14_paper_brauer.pdf

Buneman P, Khanna S, Tan WC (2001) Why and where: a characterization of data provenance. In: den Bussche JV, Vianu V (eds) Database theory – ICDT 2001. Springer, Heidelberg, pp 316–330. https://doi.org/10.1007/3-540-44503-X_20

Chapman AP, Jagadish HV (2009) Why not? In: Proceedings of the 2009 ACM SIGMOD international conference on management of data. ACM, New York, pp 523–534. https://doi.org/10.1145/1559845.1559901

Cheney J, Chiticariu L, Tan WC (2009) Provenance in databases: why, how, and where. Found Trends Databases 1(4):379–474. https://doi.org/10.1561/1900000006

Demšar J, Curk T, Erjavec A, Črt Gorup, Hočevar T, Milutinovič M, Možina M, Polajnar M, Toplak M, Starič A, Štajdohar M, Umek L, Žagar L, Žbontar J, Žitnik M, Zupan B (2013) Orange: data mining toolbox in Python. J Mach Learn Res 14(35):2349–2353. http://jmlr.org/papers/v14/demsar13a.html

Feldman M, Friedler SA, Moeller J, Scheidegger C, Venkatasubramanian S (2015) Certifying and removing disparate impact. In: Proceedings of the 21th ACM SIGKDD international conference on knowledge discovery and data mining. ACM, New York, pp 259–268. https://doi.org/10.1145/2783258.2783311

Freire J, Koop D, Santos E, Silva CT (2008) Provenance for computational tasks: a survey. Comput Sci Eng 10(3):11–21. https://doi.org/10.1109/MCSE.2008.79

Frew J, Metzger D, Slaughter P (2008) Automatic capture and reconstruction of computational provenance. Concurr Comput: Pract Exp 20(5):485–496. https://doi.org/10.1002/cpe.1247

Glavic B, Dittrich KR (2007) Data provenance: a categorization of existing approaches. In: 12. Fachtagung des GI-Fachbereichs "Datenbanken und Informationssysteme", University of Zurich, Zurich, pp 227–241. https://doi.org/10.5167/uzh-24450

Green TJ, Karvounarakis G, Tannen V (2007) Provenance semirings. In: Proceedings of the twenty-sixth ACM SIGMOD-SIGACT-SIGART symposium on principles of database systems. ACM, New York, pp 31–40. https://doi.org/10.1145/1265530.1265535

Green TJ, Karvounarakis G, Ives ZG, Tannen V (2010) Provenance in ORCHESTRA. IEEE Data Eng Bull 33(3):9–16. http://sites.computer.org/debull/A10sept/green.pdf

Guedes T, Silva V, Mattoso M, Bedo MVN, de Oliveira D (2018) A practical roadmap for provenance capture and data analysis in Spark-based scientific workflows. In: 2018 IEEE/ACM Workflows in Support of Large-Scale Science, IEEE, pp 31–41. https://doi.org/10.1109/WORKS.2018.00009

Herschel M, Diestelkämper R, Lahmar HB (2017) A survey on provenance: what for? what form? what from? VLDB J 26:881–906. https://doi.org/10.1007/s00778-017-0486-1

Huynh TD, Ebden M, Fischer J, Roberts S, Moreau L (2018) Provenance network analytics: an approach to data analytics using data provenance. Data Mining Knowl Discov 32:708–735. https://doi.org/10.1007/s10618-017-0549-3

Ikeda R, Cho J, Fang C, Salihoglu S, Torikai S, Widom J (2012) Provenance-based debugging and drill-down in data-oriented workflows. In: 28th international conference on data engineering, IEEE, Los Alamitos, CA, USA, pp 1–2. https://doi.org/10.1109/ICDE.2012.118

Interlandi M, Shah K, Tetali SD, Gulzar MA, Yoo S, Kim M, Millstein T, Condie T (2015) Titian: data provenance support in Spark. In: Proceedings of the 42nd international conference on very large data bases, pp 216–227. http://www.vldb.org/pvldb/vol9/p216-interlandi.pdf

Jentzsch SF, Hochgeschwender N (2019) Don't forget your roots! Using provenance data for transparent and explainable development of machine learning models. In: 34th IEEE/ACM international conference on automated software engineering workshop, IEEE, Los Alamitos, CA, USA, pp 37–40. https://doi.org/10.1109/ASEW.2019.00025

Koop D, Scheidegger CE, Callahan SP, Freire J, Silva CT (2008) VisComplete: automating suggestions for visualization pipelines. IEEE Trans Visual Comput Graph 14(6):1691–1698. https://doi.org/10.1109/TVCG.2008.174

Lerner BS, Boose E, Perez L (2018) Using introspection to collect provenance in R. Informatics 5(1). https://doi.org/10.3390/informatics5010012

Lim C, Lu S, Chebotko A, Fotouhi F (2010) Prospective and retrospective provenance collection in scientific workflow environments. In: 2010 IEEE international conference on services computing, IEEE, Los Alamitos, CA, USA, pp 449–456. https://doi.org/10.1109/SCC.2010.18

Ma S, Aafer Y, Xu Z, Lee WC, Zhai J, Liu Y, Zhang X (2017) LAMP: data provenance for graph-based machine learning algorithms through derivative computation. In: Proceedings of the 11th joint meeting on foundations of software engineering. ACM, New York, pp 786–797. https://doi.org/10.1145/3106237.3106291

McPhillips T, Song T, Kolisnik T, Aulenbach S, Belhajjame K, Bocinsky K, Cao Y, Chirigati F, Dey S, Freire J, Huntzinger D, Jones C, Koop D, Missier P, Schildhauer M, Schwalm C, Wei Y, Cheney J, Bieda M, Ludäscher B (2015) YesWorkflow: a user-oriented, language-independent tool for recovering workflow information from scripts. https://arxiv.org/pdf/1502.02403.pdf

Missier P, Goble C (2011) Workflows to open provenance graphs, round-trip. Fut Gener Comput Syst 27(6):812–819. https://doi.org/10.1016/j.future.2010.10.012

Mor (2013a) Constraints of the PROV data model. http://www.w3.org/TR/2013/REC-prov-constraints-20130430/

Mor (2013b) PROV-DM: the PROV data model. https://www.w3.org/TR/prov-dm/

Murta L, Braganholo V, Chirigati F, Koop D, Freire J (2015) noWorkflow: capturing and analyzing provenance of scripts. In: Ludäscher B, Plale B (eds) Provenance and annotation of data and processes. Springer, Cham, pp 71–83. https://doi.org/10.1007/978-3-319-16462-5_6

Namaki MH, Floratou A, Psallidas F, Krishnan S, Agrawal A, Wu Y (2020) Vamsa: tracking provenance in data science scripts. https://arxiv.org/pdf/2001.01861.pdf

Packer HS, Chapman A, Carr L (2019) GitHub2PROV: provenance for supporting software project management. In: 11th international workshop on theory and practice of provenance. https://www.usenix.org/system/files/tapp2019-paper-packer.pdf

Pimentel JF, Dey S, McPhillips T, Belhajjame K, Koop D, Murta L, Braganholo V, Ludäscher B (2016a) Yin & Yang: demonstrating complementary provenance from noWorkflow & YesWorkflow. In: Mattoso M, Glavic B (eds) Provenance and annotation of data and processes. Springer, Cham, pp 161–165. https://doi.org/10.1007/978-3-319-40593-3_13

Pimentel JF, Freire J, Murta L, Braganholo V (2016b) Fine-grained provenance collection over scripts through program slicing. In: Mattoso M, Glavic B (eds) Provenance and annotation of data and processes. Springer, Cham, pp 199–203. https://doi.org/10.1007/978-3-319-40593-3_21

Pimentel JF, Murta L, Braganholo V, Freire J (2017) noWorkflow: a tool for collecting, analyzing, and managing provenance from Python scripts. Proc VLDB Endowm 10(12):1841–1844. https://doi.org/10.14778/3137765.3137789

Pimentel JF, Freire J, Murta L, Braganholo V (2019) A survey on collecting, managing, and analyzing provenance from scripts. ACM Comput Surv 52(3). https://doi.org/10.1145/3311955

Psallidas F, Wu E (2018) Provenance for interactive visualizations. In: Proceedings of the workshop on human-in-the-loop data analytics. ACM, New York. https://doi.org/10.1145/3209900.3209904

Roper B, Chapman A, Martin D, Cavazzi S (2020) Mapping trusted paths to VGI. ProvenanceWeek 2020, virtual event, poster

Santos E, Koop D, Vo HT, Anderson EW, Freire J, Silva C (2009) Using workflow medleys to streamline exploratory tasks. In: Winslett M (ed) Scientific and statistical database management. Springer, Heidelberg, pp 292–301. https://doi.org/10.1007/978-3-642-02279-1_23

Sarikhani M, Wendelborn A (2018) Mechanisms for provenance collection in scientific workflow systems. Computing 100:439–472. https://doi.org/10.1007/s00607-017-0578-1

Sasikant A (2019) Provenance capture mechanism for Orange, a data mining and machine learning toolkit, to evaluate the effectiveness of provenance capture in machine learning. Thesis, University of Southampton, Southampton

Shang Z, Zgraggen E, Buratti B, Kossmann F, Eichmann P, Chung Y, Binnig C, Upfal E, Kraska T (2019) Democratizing data science through interactive curation of ML pipelines. In: Proceedings of the 2019 international conference on management of data. ACM, New York, pp 1171–1188. https://doi.org/10.1145/3299869.3319863

Simmhan YL, Plale B, Gannon D (2005) A survey of data provenance in e-Science. ACM SIGMOD Record 34(3):31–36. https://doi.org/10.1145/1084805.1084812

Simonelli G (2019) Capturing and querying fine-grained provenance of preprocessing pipelines in data science. Thesis, Università Roma Tre, Rome

Souza R, Silva V, Coutinho ALGA, Valduriez P, Mattoso M (2017) Data reduction in scientific workflows using provenance monitoring and user steering. Fut Gener Comput Syst 110:481–501. https://doi.org/10.1016/j.future.2017.11.028

Tang M, Shao S, Yang W, Liang Y, Yu Y, Saha B, Hyun D (2019) SAC: a system for Big Data lineage tracking. In: 35th international conference on data engineering, IEEE, pp 1–2. https://doi.org/10.1109/ICDE.2019.00215

Thavasimani P, Caa J, Missier P (2019) Why-diff: exploiting provenance to understand outcome differences from non-identical reproduced workflows. IEEE Access 7:34973–34990. https://doi.org/10.1109/ACCESS.2019.2903727

Zelaya CVG (2019) Towards explaining the effects of data preprocessing on machine learning. In: 35th international conference on data engineering, IEEE, pp 2086–2090. https://doi.org/10.1109/ICDE.2019.00245

Zelaya VG, Missier P, Prangle D (2019) Parametrised data sampling for fairness optimisation. Explainable AI for fairness, accountability & transparency workshop, Anchorage, AK

Zhang Q, Morris PJ, McPhillips T, Hanken J, Lowery DB, Ludäscher B, Macklin JA, Morris RA, Wieczorek J (2017) Using YesWorkflow hybrid queries to reveal data lineage from data curation activities. Biodivers Inf Sci Stand 1:e20380. https://doi.org/10.3897/tdwgproceedings.1.20380

Chapter 4
Contextualized Knowledge Graphs in Communication Network and Cyber-Physical System Modeling

Leslie F. Sikos ⓘ

4.1 Introduction

There is a range of challenges software agents have to face when processing networking data and cyberthreat intelligence, from data heterogeneity to a lack of consensus in network terminology. The following sections briefly discuss these.

4.1.1 Heterogeneity Issues of Representing Cyber-Knowledge

The enormous variety of network infrastructures and computing platforms causes challenges for processing cyber-knowledge efficiently and automatically. The lack of consensus and vendor-specific data serializations on networking devices cause data heterogeneity issues, which software agents often cannot handle. For example, the naming convention of Cisco routers are different from that of Juniper routers, making it challenging or infeasible to capture the semantics of networking concepts of such devices without having a machine-interpretable definition that also covers the equivalence of two concepts using different names. If an acronym of a concept (which is commonplace in networking) is used instead of writing it out in full, matching identical concepts and/or individuals can be problematic.

Similarly, the connection between concepts such as neighbors and adjacency is not machine-interpretable unless there is a definition in a knowledge organization system, such as an ontology, for this. Think about how this affects the work of network administrators who specialize in one system or the other, and even those who can work with both have to use different commands depending on the system

L. F. Sikos (✉)
Edith Cowan University, Perth, WA, Australia
e-mail: l.sikos@ecu.edu.au

© Springer Nature Switzerland AG 2021
L. F. Sikos et al. (eds.), *Provenance in Data Science*, Advanced Information and Knowledge Processing, https://doi.org/10.1007/978-3-030-67681-0_4

used. For example, the `show clns neighbors` command in Cisco IOS has the equivalent `show isis adjacency` in Juniper's Junos OS for summarizing known neighbors, the connecting interface, and the state of adjacency; however, the correlation between neighbors and adjacency is not necessarily obvious for a software agent.

On top of the complexities of communication networks, smart devices and IoT devices often use their proprietary data formats, which have to be converted from one to another, or represented uniformly to allow automation in intrusion detection (Sikos 2020b), incident response, and network forensics (Sikos 2020a).

4.1.2   Introduction to Knowledge Graphs in Cybersecurity Applications

The ever-growing and vast amount of information on the Web requires sophisticated indexing and search mechanisms to facilitate efficient processing, such as via automated reasoning (Sikos 2016). These are supported by a variety of knowledge representations, such as *knowledge graphs*,[1] which are graph-based representations that integrate data about knowledge domain entities, interconnected via machine-interpretable relations (Sikos et al. 2018e). Knowledge graphs enable graph algorithms, such as graph traversal, to be applied, thereby facilitating alternate data processing options, knowledge fusion (Qi et al. 2020), and advanced querying, for example, to detect potential cyberthreats via pattern matching (Chowdhury et al. 2017). Knowledge graphs can be utilized in a variety of information processing and management tasks, from semantic search, recommendations, and summarization to data integration, empowering ML and NLP techniques, and improving automation and robotics (Sheth et al. 2020).

To ensure the machine-interpretability of, and enable automated reasoning over, network data such as packet captures, a structured data format and uniform representation are needed (Sikos 2019, 2020c), which form a fundamental part of knowledge graphs representing cyber-knowledge. Such knowledge graphs can be stored in graph databases based either on the Resource Description Framework (RDF) standard (i.e., triplestores and quadstores), or on labeled property graphs (popularized by Neo4j[2]), which are among the most dominant graph data models besides attributed graph (Akoglu et al. 2014), hypergraph (Guzzo et al. 2014; Jagtap and Sriram 2019), and conceptual graph (Chein and Mugnier 2015).

---

[1] The term was popularized by Google when introducing their proprietary knowledge base, *Google Knowledge Graph*, in 2012 to enhance the value of information returned by Google Web Search queries based on aggregating data from unstructured, semistructured, and structured data sources (Sikos 2015).

[2] https://neo4j.com

Contextualized knowledge graphs are knowledge graphs that represent statements complemented by context. A large share of communication network data represented using knowledge graphs cannot be used for verification purposes and would not be considered authoritative and reliable unless the context is captured with each graph edge (corresponding to a property or relation). For example, to assign reliability weights to network knowledge statements derived from diverse sources, the data source has to be captured—think of determining which one is more trustworthy: information from a router configuration file set up by an organization's system administrator, or computer-generated information from an OSPF Link State Advertisement. Capturing provenance of communication networks can also indicate how up-to-date is the represented cyber-knowledge, such as by considering the update frequency of routing messages on a network.

How provenance can be captured in a knowledge graph depends on the graph data model used. RDF-based cyber-knowledge graphs, for example, typically use RDF quadruples (RDF quads), often implemented as named RDF graphs, to capture context (Sikos and Philp 2020). This makes it possible to make realistic judgments for cyber-situational awareness even if some of the network knowledge statements are contradictory (for example, if one statement is more up-to-date than the other, or came from a more trusted source).

4.2 Utilizing Cybersecurity Ontologies in Knowledge Graphs

Cybersecurity—and some network—ontologies can be used to capture cyber-knowledge in knowledge bases and facilitate automated reasoning to obtain cyber-situational awareness and support decision-making when detecting cyberthreats based on complex behavioral and attack patterns (Sikos 2018a). Some prime examples for utilizing ontologies to represent cyber-knowledge in graphs are summarized below.

Iannacone et al. (2015) developed the *Situation and Threat Understanding by Correlating Contextual Observations (STUCCO) Ontology*[3] for cybersecurity knowledge graphs, which formally defines the relationship between concepts such as user, account, host, host address, IP, address range, port, service, domain name, software, vulnerability, malware, attack, flow, and attacker.

Chabot et al. (2015) introduced an approach called *Semantic Analysis of Digital Forensic Cases (SADFC)* based on an ontology, *ORD2I*, to represent digital events with digital forensic considerations, and in particular, repeatability, in mind. Knowledge graphs about digital events can be generated by instantiating ORD2I concepts. The formal representation of digital events using this approach enables automated reasoning, efficient querying with SPARQL, and visualizations in the form of correlation graphs.

---

[3]https://stucco.github.io

Jia et al. (2018) proposed a quintuple-based cybersecurity knowledge base model that covers five aspects: concept, instance, relation, properties, and rule. As part of the model, three ontologies are used for assets, vulnerabilities, and attacks, respectively.

The *Communication Network Topology and Forwarding Ontology (CNTFO)*[4] was designed to capture the semantics of communication network layout and information flow (Sikos et al. 2018d). This can be used to create network security datasets like *ISPNet*,[5] which are based on contextualized knowledge graphs of autonomous systems and the network devices interconnected in them, complemented by their properties bound by precise, standard-aligned datatypes and property value ranges (Sikos et al. 2018a). Knowledge representations utilizing CNTFO can be complemented by provenance data, such as via named graphs.

Using a threat ontology and a knowledge graph, Gong and Tian (2020) specified a knowledge base and unified description as part of an approach for threat modeling. These can be used to establish adaptive security systems that are capable of analyzing potential cyberthreats and providing solutions for security architectures, and can help obtain cyber-situational awareness and facilitate intelligent decision-making.

Grangel-González et al. (2018) integrated various perspectives of cyber-physical systems into a knowledge graph in which each fact is assigned a weight in the range [0,1] to represent fact certainty. This knowledge graph is formally defined as a quintuple of the form $\mathcal{G}_u = \langle I, V, L, D, U \rangle$, where I and V are sets that correspond to URIs that identify elements in a cyber-physical system's document and terms from a cyber-physical system standard vocabulary, L is a set of literals, D is an RDF graph of the form $(s, p, o) \in I \times V \times (I \cup L)$ representing hard knowledge facts, and U is an RDF graph of the form $U = \{(t, w) | t \in I \times V \times (I \cup L) \text{ and } w \in [0, 1]\}$, i.e., weights are assigned to each triple. Such knowledge graphs can be used to represent soft and hard knowledge facts from cyber-physical system descriptions.

Asamoah et al. (2016) proposed an approach that utilizes RDF/XML-serialized knowledge graphs for the filtration process of cloud-based cybersecurity ecosystems. This approach consists of an algorithm for classifying and filtering cyberevents, and an extension to the standard OWL to support custom relationships.

4.3   Graph-Based Visualization

Knowledge graph-based cybersecurity visualizations are on the rise, as evidenced by systems such as *CyGraph*, which can help maintain cyber-situational awareness and protect mission-critical assets by capturing complex relationships between cybersecurity entities (Noel et al. 2016). It covers incremental attack vulnerability,

---

[4]https://purl.org/ontology/network/

[5]https://purl.org/dataset/ispnet/base/

security events, and mission dependencies of communications networks, builds a predictive model of possible attack paths and critical vulnerabilities, and correlates events to known vulnerability paths.

Cyberthreat intelligence can benefit from graph-based visualizations, as seen in the example of *KAVAS (Knowledge-Assisted Visual Analytics for STIX)* (Böhm et al. 2018). KAVAS utilizes the Neo4j graph database for storing and visualizing cyberknowledge graphs that instantiate STIX objects, such as attack patterns and threat actors.

Graphs are inherently powerful in displaying network devices in communication networks, such as routers and switches, as well as the applications hosted on these, where the graph edges can indicate the connection between the network devices, such as Cat6 cables. Graphs can also be considered a good choice when modeling how malware propagates across different network devices. Insider threats, which are notorious for being difficult to detect, can be efficiently indicated by cyberthreat graphs, and determine who might have had access to information and how they might have used it maliciously. *EclecticIQ*[6] provides assistance to understand complex and disparate cyberthreat intelligence data by combining graph technologies, including advanced graph search and network graph correlation matrices. The resulting visualization can transform work-intensive manual data exploration into a fast extraction and interpretation process via an intuitive interface while communicating critical insight, as for example, about a threat actor in relation to IDS data (see Fig. 4.1).

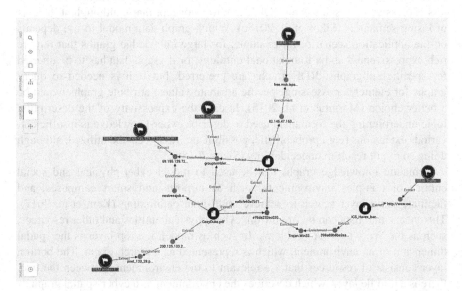

Fig. 4.1 Graph-based visualization of cyberthreat intelligence data in EclecticIQ.[7]

---

[6]https://www.eclecticiq.com
[7]https://youtu.be/14OPKlBIt5s?t=1559

EclecticIQ graphs can show IP communications with individual IP addresses and transfer protocols, where the edges represent the communication data, such as an acknowledgment (ACK) signal, and complex relationships between network devices, users, and security information and event management (SIEM). Aligned with STIX and TAXII, the threat data in EclecticIQ can be serialized in JSON for further processing in other applications.

The Graph Diagram widget of Devo[8] (formerly Logtrust) can visualize the relationship between nodes of a given type, such as the relationship between source and destination IPs in a web traffic log file. In addition, if location information is provided to each of the nodes, the graph can be superimposed over a map to get a graphical representation adding the location value to the nodes' relationship.

4.4 Using Knowledge Graphs in Security System and Cyber-Physical System Modeling

The various graph data models have different strengths and weaknesses, for example, RDF graphs have URI identifiers and rich semantics (Sikos 2017), but require technical knowledge; labeled property graphs have an internal structure and allow multiple property value pairs per edge (Robinson et al. 2015), but are proprietary. Some of these are more expressive than others, meaning that a fallback mechanism can be realized when reducing a more expressive representation to a less expressive one (such as an RDF graph to a concept map), although at the cost of losing semantics (Sikos et al. 2018c). Which graph data model to use depends on the application scenario; for example, for large knowledge graphs that require rich expressiveness and a foundational ontology, or if graph data has to be queried for specific subgraphs, RDF graphs are preferred, but if it is needed to query graphs for elements possessing specific attribute values, attribute graphs would be a better choice (Margitus et al. 2015). In case the expressivity of the description logic underpinning the formalism used to describe cyber-knowledge is insufficient, various extensions (e.g., probabilistic, possibilistic, fuzzy) can be utilized, although doing so might result in undecidability (Sikos 2018b).

Semantic knowledge graphs can be used to model cyber-physical and social entities of complex environments, such as airports and smart campuses, and facilitate multi-layer search to achieve multi-goal pathfinding (Kem et al. 2017). This means that data can be fused from cyber-physical entities and other resources, such as the Web, to solve problems. In such as model, the top layer is the spatial dimension of an environment, which is represented as a search graph. The bottom layer consists of resources that are relevant to the environment. Between the two, there is a middle layer, which describes the environment as a cyber-spatial graph.

[8]https://www.devo.com

The automation of the design of cyber-physical systems can be supported by machine learning, such as via graph learning using convolutional neural networks (Chhetri et al. 2019); however, choosing a suitable type of graph representation is crucial. This is because modeling cyber-physical systems with graph-based models requires graph arcs that are not limited to an association with predicates or properties, as seen in RDF graphs, and have to support graph-theoretical algorithms for analyzing the overall graph structure that goes beyond the capabilities of RDF graphs (Privat and Abbas 2019). Some of the semantics of cyber-physical systems apply to the graph as a whole rather than the per resource semantics RDF triples are meant to describe. Therefore, property graphs can be considered better metamodels for capturing the semantics of cyber-physical systems. Other purpose-designed non-RDF graphs have also been proposed, such as Web of Things (WoT) graphs (Privat et al. 2019).

4.5 Task Automation in Cyberthreat Intelligence and Cyber-Situational Awareness Using Knowledge Graphs

In threat intelligence recommendation systems, generic knowledge graphs often hold too much irrelevant data and cannot consider entity relationships in terms of attack chains, which can be addressed by providing threat intelligence not only for machines, but also for human consumption (Du et al. 2019). Nevertheless, the formal representation of cyberthreats can facilitate data-driven applications in cyberthreat intelligence via automated reasoning (Sikos and Choo 2020). Creating knowledge graphs for communication networks enables automated reasoning to achieve cyber-situational awareness, such as by automatically pinpointing relationships between seemingly unrelated network entities, which is particularly useful if there are several network segments between them, or if they are under the management of different autonomous systems (Sikos et al. 2018b). The uniform representation of network knowledge in RDF can utilize the standard RDFS and OWL entailment regimes, making it possible to infer new statements by making implicit knowledge explicit. Such statements can be displayed on dashboards for decision support, and can vastly improve the detection of duplicate network entities having multiple names,[9] and correlations even experienced network analysts would not notice.

Alqahtani et al. (2016) introduced a modeling approach that utilizes Semantic Web technologies to establish bi-directional traceability links between security advisory repositories and other software repositories via a unified ontological representation. By performing semantic inference, both direct and transitive depen-

---

[9]This is common due to (1) the different naming conventions used by the various data sources, such as routing messages and router configuration files, and (2) some graph nodes are initially blank because of the unavailability of the proper/descriptive name.

dencies can be determined between reported vulnerabilities and potentially affected Maven projects, thereby highlighting the potential impact of vulnerable component reuse in a global software ecosystem context.

By fusing knowledge graphs representing concepts and properties of the industry standard *Common Attack Pattern Enumeration and Classification (CAPEC)*,[10] *Common Vulnerabilities and Exposures (CVE)*,[11] *National Vulnerability Database (NVD)*,[12] *Common Vulnerability Scoring System (CVSS)*,[13] and *Common Weakness Enumeration Specification (CWE)*,[14] one-to-many relationships between CAPEC and CVE entities[15] can be formalized through CWE, which enables alert correlation analysis (Wang et al. 2017).

Kiesling et al. (2019) constructed a knowledge graph for security event streams that covers vulnerabilities and attack patterns from public resources, aligned with CVSS, CWE, and CAPEC concepts. The architecture of the graph was designed with constant updates in mind and provides graph data exposure as Linked Data, triple pattern fragments,[16] RDF dumps,[17] and via a SPARQL endpoint.[18] This knowledge graph can be used in security applications from vulnerability assessment to intrusion detection.

By fusing ontology-captured network knowledge about topology, services, vulnerabilities, and configurations, attack graphs can be generated iteratively and potential attacks inferred via automated reasoning (Wu et al. 2017). Structured data representations of security logs can be enriched with semantics to facilitate unambiguous and useful interlinking between logs in knowledge graphs (Kurniawan 2018). By performing reasoning over such representations, suspicious behaviors and potential attacks can be identified in real time via the utilization of property chains and the transitive and reflective properties of interlinked security log data.

Contextualized reasoning over RDF-quad-based cyber-knowledge graphs can be used for decision support (Philp et al. 2019b). The entities identified using names derived from diverse data sources or using an initial/random name used in a number of RDF statements can reveal different properties and facts of dynamic communication network knowledge. Without identifying multiple representations of the same real-world entity, cyber-knowledge remains fragmented as separate branches of the same graph or as different graphs, which prevents deduction based on the correlations between the objects of statements in which the duplicate entity is the subject. By facilitating entity resolution on such graphs, richer semantics can

---

[10] https://capec.mitre.org

[11] https://cve.mitre.org

[12] https://nvd.nist.gov

[13] https://www.first.org/cvss/v3.1/specification-document

[14] https://cwe.mitre.org

[15] This is because there are many vulnerabilities related to each attack type.

[16] https://ldf-server.sepses.ifs.tuwien.ac.at/

[17] https://sepses.ifs.tuwien.ac.at/dumps/

[18] https://sepses.ifs.tuwien.ac.at/sparql

be captured in the representation, which can be used for querying, reasoning, and data mining (Philp et al. 2019a).

Cyber-knowledge graphs complemented by a set of rules can facilitate cyber-situational awareness via rule mining, as seen with the *Subsumption Reasoning for Rule Deduction (SRRD)* approach of Liu et al. (2020). This approach utilizes background knowledge represented by first-order logic to obtain Horn rules from cyber-knowledge graphs through a series of pruning and query rewriting techniques. By automatically discovering redundant semantic association rules based on knowledge graph reasoning, redundant rules can be reduced while general, comprehensive, and streamlined rules can be preserved, thereby reducing the workload of security analysts.

4.6 Summary

Because of the heterogeneous data typical to communication networks and ubiquitous computing, the automation of processing cyber-knowledge is challenging. A promising research direction in this field is to employ formal knowledge representation, and knowledge graphs in particular, to provide a uniform representation for data derived from diverse sources. Knowledge graphs are widely adopted for capturing the semantics of, and visualizing, cyber-knowledge in cybersecurity, cybersituational awareness, and cyberthreat intelligence. Many of these are RDF-based and use purpose-designed cybersecurity ontologies to formally represent cyberthreats, cyber-events, and cyber-actors. Not RDF-based knowledge graphs, such as cyber-physical graphs and WoT graphs, can also be used in security applications, and can be more efficient in modeling certain aspects of cyber-physical systems than RDF graphs.

References

Akoglu L, Tong H, Koutra D (2014) Graph based anomaly detection and description: a survey. Data Min Knowl Disc 29(3):626–688. https://doi.org/10.1007/s10618-014-0365-y

Alqahtani SS, Eghan EE, Rilling J (2016) Tracing known security vulnerabilities in software repositories—a Semantic Web enabled modeling approach. Sci Comput Program 121:153–175. https://doi.org/10.1016/j.scico.2016.01.005

Asamoah C, Tao L, Gai K, Jiang N (2016) Powering filtration process of cyber security ecosystem using knowledge graph. In: Qiu M, Tao L, Niu J (eds) Proceedings of the 3rd International Conference on Cyber Security and Cloud Computing (CSCloud). IEEE Computer Society, Los Alamitos, pp 240–246. https://doi.org/10.1109/CSCloud.2016.36

Böhm F, Menges F, Pernul G (2018) Graph-based visual analytics for cyber threat intelligence. Cybersecurity 1(1), 16. https://doi.org/10.1186/s42400-018-0017-4

Chabot Y, Bertaux A, Nicolle C, Kechadi T (2015) An ontology-based approach for the reconstruction and analysis of digital incidents timelines. Digit Investig 15:83–100. https://doi.org/10.1016/j.diin.2015.07.005

Chein M, Mugnier ML (2015) Graph-based knowledge representation: computational foundations of conceptual graphs. Springer, London. https://doi.org/10.1007/978-1-84800-286-9

Chhetri SR, Wan J, Canedo A, Faruque MAA (2019) Design automation using structural graph convolutional neural networks. In: Faruque MAA, Canedo A (eds) Design automation of cyber-physical systems. Springer, Cham, chap 9, pp 237–259. https://doi.org/10.1007/978-3-030-13050-3_9

Chowdhury FARR, Ma C, Islam MR, Namaki MH, Faruk MO, Doppa JR (2017) Select-and-evaluate: a learning framework for large-scale knowledge graph search. In: Zhang ML, Noh YK (eds) Proceedings of machine learning research. PMLR, Cambridge, vol 77, pp 129–144. http://proceedings.mlr.press/v77/chowdhury17a/chowdhury17a.pdf

Du M, Jiang J, Jiang Z, Lu Z, Du X (2019) PRTIRG: a knowledge graph for people-readable threat intelligence recommendation. In: Douligeris C, Karagiannis D, Apostolou D (eds) Knowledge science, engineering and management. Springer, Cham, pp 47–59. https://doi.org/10.1007/978-3-030-29551-6_5

Gong L, Tian Y (2020) Threat modeling for cyber range: an ontology-based approach. In: Liang Q, Liu X, Na Z, Wang W, Mu J, Zhang B (eds) Communications, signal processing, and systems. Springer, Singapore, pp 1055–1062. https://doi.org/10.1007/978-981-13-6508-9_128

Grangel-González I, Halilaj L, Vidal ME, Rana O, Lohmann S, Auer S, Müller AW (2018) Knowledge graphs for semantically integrating cyber-physical systems. In: Hartmann S, Ma H, Hameurlain A, Pernul G, Wagner RR (eds) Database and expert systems applications. Springer, Cham, pp 184–199. https://doi.org/10.1007/978-3-319-98809-2_12

Guzzo A, Pugliese A, Rullo A, Saccà D (2014) Intrusion detection with hypergraph-based attack models. In: Croitoru M, Rudolph S, Woltran S, Gonzales C (eds) Graph structures for knowledge representation and reasoning. Springer, Cham, pp 58–73. https://doi.org/10.1007/978-3-319-04534-4_5

Iannacone M, Bohn S, Nakamura G, Gerth J, Huffer K, Bridges R, Ferragut E, Goodall J (2015) Developing an ontology for cyber security knowledge graphs. In: Trien JP (ed) Proceedings of the 10th Annual Cyber and Information Security Research Conference, ACM, New York. https://doi.org/10.1145/2746266.2746278

Jagtap SS, Sriram VSS (2019) Subtree hypergraph-based attack detection model for signature matching over SCADA HMI. In: Sriram VSS, Subramaniyaswamy V, Sasikaladevi N, Zhang L, Batten L, Li G (eds) Applications and techniques in information security. Springer, Singapore, pp 173–184. https://doi.org/10.1007/978-981-15-0871-4_13

Jia Y, Qi Y, Shang H, Jiang R, Li A (2018) A practical approach to constructing a knowledge graph for cybersecurity. Engineering 4(1):53–60. https://doi.org/10.1016/j.eng.2018.01.004

Kem O, Balbo F, Zimmermann A, Nagellen P (2017) Multi-goal pathfinding in cyber-physical-social environments: Multi-layer search over a semantic knowledge graph. Procedia Comput Sci 112:741–750. https://doi.org/10.1016/j.procs.2017.08.162

Kiesling E, Ekelhart A, Kurniawan K, Ekaputra F (2019) The SEPSES knowledge graph: an integrated resource for cybersecurity. In: Ghidini C, Hartig O, Maleshkova M, Svátek V, Cruz I, Hogan A, Song J, Lefrançois M, Gandon F (eds) The Semantic Web—ISWC 2019. Springer, Cham, pp 198–214. https://doi.org/10.1007/978-3-030-30796-7_13

Kurniawan K (2018) Semantic query federation for scalable security log analysis. In: Gangemi A, Gentile AL, Nuzzolese AG, Rudolph S, Maleshkova M, Paulheim H, Pan JZ, Alam M (eds) The Semantic Web: ESWC 2018 satellite events. Springer, Cham, pp 294–303. https://doi.org/10.1007/978-3-319-98192-5_48

Liu B, Zhu X, Wu J, Yao L (2020) Rule reduction after knowledge graph mining for cyber situational awareness analysis. Procedia Comput Sci 176:22–30. https://doi.org/10.1016/j.procs.2020.08.003

Margitus M, Tauer G, Sudit M (2015) RDF versus attributed graphs: the war for the best graph representation. In: Proceedings of the 18th International Conference on Information Fusion. IEEE, New York, pp 200–2006

Noel S, Harley E, Tam KH, Limiero M, Share M (2016) CyGraph: graph-based analytics and visualization for cybersecurity. In: Gudivada VN, Raghavan VV, Govindaraju V, Rao CR (eds) Cognitive computing: theory and applications, chap 4, pp 117–167. https://doi.org/10.1016/bs.host.2016.07.001

Philp D, Chan N, Mayer W (2019a) Network path estimation in uncertain data via entity resolution. In: Le TD, Ong KL, Zhao Y, Jin WH, Wong S, Liu L, Williams G (eds) Data mining. Springer, Singapore, pp 196–207. https://doi.org/10.1007/978-981-15-1699-3_16

Philp D, Chan N, Sikos LF (2019b) Decision support for network path estimation via automated reasoning. In: Czarnowski I, Howlett RJ, Jain LC (eds) Intelligent decision technologies 2019. Springer, Singapore, pp 335–344. https://doi.org/10.1007/978-981-13-8311-3_29

Privat G, Abbas A (2019) "Cyber-Physical graphs" vs. RDF graphs. https://www.w3.org/Data/events/data-ws-2019/assets/position/Gilles%20Privat.html. W3C Workshop on Web Standardization for Graph Data

Privat G, Coupaye T, Bolle S, Raipin-Parvedy P (2019) WoT graph as multiscale digital-twin for cyber-physical systems-of-systems. https://www.w3.org/WoT/ws-2019/Presentations%20-%20Day%202/Future%20Work/10_WoT%20Graph%20as%20Multiscale%20Digital-Twin_2019-06-05_WoT_G.Privat.pdf. 2nd W3C Web of Things Workshop, Munich, Germany

Qi G, Chen H, Liu K, Wang H, Ji Q, Wu T (2020) Knowledge graph. Springer, Singapore

Robinson I, Webber J, Eifrem E (2015) The labeled property graph model. In: Graph databases: new opportunities for connected data, 2nd edn. O'Reilly Media, New York

Sheth A, Padhee S, Gyrard A (2020) Knowledge graphs and knowledge networks: the story in brief. IEEE Internet Comput 23:67–75. https://doi.org/10.1109/MIC.2019.2928449

Sikos LF (2015) Google knowledge graph and knowledge vault, in: Mastering structured data on the Semantic Web Apress, pp 200–205. https://doi.org/10.1007/978-1-4842-1049-9_8

Sikos LF (2016) A novel approach to multimedia ontology engineering for automated reasoning over audiovisual LOD datasets. In: Nguyen NT, Trawiski B, Fujita H, Hong TP (eds) Intelligent information and database systems. Springer, Heidelberg, pp 3–12. https://doi.org/10.1007/978-3-662-49381-6_1

Sikos LF (2017) Description logics in multimedia reasoning. Springer, Cham. https://doi.org/10.1007/978-3-319-54066-5

Sikos LF (ed) (2018a) AI in cybersecurity. Springer, Cham. https://doi.org/10.1007/978-3-319-98842-9

Sikos LF (2018b) Handling uncertainty and vagueness in network knowledge representation for cyberthreat intelligence. In: Proceedings of the 2018 IEEE International Conference on Fuzzy Systems. IEEE, New York. https://doi.org/10.1109/FUZZ-IEEE.2018.8491686

Sikos LF (2019) Knowledge representation to support partially automated honeypot analysis based on Wireshark packet capture files. In: Czarnowski I, Howlett RJ, Jain LC (eds) Intelligent Decision Technologies 2019. Springer, Singapore, pp 335–344. https://doi.org/10.1007/978-981-13-8311-3_30

Sikos LF (2020a) AI in digital forensics: ontology engineering for cybercrime investigations. WIREs Forensic Science, p e1394. https://doi.org/10.1002/wfs2.1394

Sikos LF (2020b) AI-powered cybersecurity: from automated threat detection to adaptive defense. CISO MAG 4(5):74–87

Sikos LF (2020c) Packet analysis for network forensics: a comprehensive survey. Forensic Sci. Int. Digit. Investig. 32C:200, 892. https://doi.org/10.1016/j.fsidi.2019.200892

Sikos LF, Choo KKR (eds) (2020) Data science in cybersecurity and cyberthreat intelligence. Springer, Cham. https://doi.org/10.1007/978-3-030-38788-4

Sikos LF, Philp D (2020) Provenance-aware knowledge representation: a survey of data models and contextualized knowledge graphs. Data Sci Eng. https://doi.org/10.1007/s41019-020-00118-0

Sikos LF, Philp D, Voigt S, Howard C, Stumptner M, Mayer W (2018a) Provenance-aware LOD datasets for detecting network inconsistencies. In: Capadisli S, Cotton F, Giménez-García JM, Haller A, Kalampokis E, Nguyen V, Sheth A, Troncy R (eds) Joint Proceedings of the International Workshops on Contextualized Knowledge Graphs, and Semantic Statistics

Co-Located with 17th International Semantic Web Conference, RWTH Aachen University, Aachen. http://ceur-ws.org/Vol-2317/article-03.pdf

Sikos LF, Stumptner M, Mayer W, Howard C, Voigt S, Philp D (2018b) Automated reasoning over provenance-aware communication network knowledge in support of cyber-situational awareness. In: Liu W, Giunchiglia F, Yang B (eds) Knowledge science, engineering and management. Springer, Cham, pp 132–143. https://doi.org/10.1007/978-3-319-99247-1_12

Sikos LF, Stumptner M, Mayer W, Howard C, Voigt S, Philp D (2018c) Representing conceptualized dynamic network knowledge for cyber-situational awareness. In: Cañas AJ, Reiska P, Zea C, Novak JD (eds) Proceedings of the 8th International Conference on Concept Mapping: Renewing Learning and Thinking, p 396

Sikos LF, Stumptner M, Mayer W, Howard C, Voigt S, Philp D (2018d) Representing network knowledge using provenance-aware formalisms for cyber-situational awareness. Procedia Comput Sci 126C:29–38. https://doi.org/10.1016/j.procs.2018.07.206

Sikos LF, Stumptner M, Mayer W, Howard C, Voigt S, Philp D (2018e) Summarizing network information for cyber-situational awareness via cyber-knowledge integration. AOC 2018 Convention, Adelaide, Australia, 30–May 2018

Wang W, Jiang R, Jia Y, Li A, Chen Y (2017) KGBIAC: knowledge graph based intelligent alert correlation framework. In: Wen S, Wu W, Castiglione A (eds) Cyberspace Safety and Security. Springer, Cham, pp 523–530. https://doi.org/10.1007/978-3-319-69471-9_41

Wu S, Zhang Y, Cao W (2017) Network security assessment using a semantic reasoning and graph-based approach. Comput Electr Eng 64:96–109. https://doi.org/10.1016/j.compeleceng.2017.02.001

Chapter 5
ProvCaRe: A Large-Scale Semantic Provenance Resource for Scientific Reproducibility

Chang Liu, Matthew Kim, Michael Rueschman, and Satya S. Sahoo ⓘD

5.1 Introduction

Reproducibility of research results is a key component of scientific research across domains (Munafò et al. 2017; Stodden et al. 2013). In particular, scientific reproducibility has become a critical concern in the biomedical and health research domains with several papers identifying a significant lack of reproducibility in research studies. For example, in cancer research, a study found only 6 out of 53 published works to be reproducible, i.e., 89% of highly rated cancer research findings could not be confirmed independently (Begley and Ellis 2012). Similarly, according to a survey of more than 1500 researchers published in the journal *Nature*, 50% of the respondents could not reproduce results from their own experiments and approximately 70% of the respondents could not reproduce results of experiments conducted by other researchers (Baker 2016). The alarming results from these and several other studies led to several initiatives to enable scientific reproducibility. In particular, the impact of non-reproducible research in the biomedical and health domains with a focus on the expensive process of new drug development, patient safety, and the quality of life of individuals has led to the development of the US National Institutes of Health (NIH) "Rigor and Reproducibility" guidelines (Collins and Tabak 2014; Landis et al. 2012).

The NIH Rigor and Reproducibility guidelines together with several other initiatives to support scientific reproducibility (Schulz et al. 2010; Kilkenny et al. 2010) are synergistic with the increasing availability of large-scale biomedical data

C. Liu (✉) · S. S. Sahoo
Case Western Reserve University, Cleveland, OH, USA
e-mail: satya.sahoo@case.edu

M. Kim · M. Rueschman
Harvard University, Boston, MA, USA
e-mail: mrueschman@bwh.harvard.edu

© Springer Nature Switzerland AG 2021   59
L. F. Sikos et al. (eds.), *Provenance in Data Science*, Advanced Information
and Knowledge Processing, https://doi.org/10.1007/978-3-030-67681-0_5

repositories (The Cancer Genome Atlas Research Network et al. 2013; NIM 2020). However, without the availability of essential contextual metadata or provenance information describing how the data in these large repositories were recorded, processed, and analyzed, it is extremely difficult to achieve the objective of scientific reproducibility. However, there are no existing provenance metadata repositories corresponding to large-scale biomedical data repositories, which can be used for reproducing the results of published articles that are available in the National Center for Biotechnology Information (NCBI) MEDLINE repository. MEDLINE is one of the largest repositories of published biomedical articles with more than 26 million records that can be queried using the PubMed query interface.

Using the NCBI MEDLINE repository, the *Provenance for Clinical and Health Research (ProvCaRe)* project[1] has been developed to support scientific reproducibility in biomedical sciences with three contributions:

1. *Development of a provenance ontology* to model a core set of concepts to support scientific reproducibility in biomedical research by extending the W3C PROV Ontology (PROV-O);
2. *Creation of a comprehensive semantic provenance repository* by identifying and extracting structured provenance information from all available published articles using NCBI MEDLINE and NCBI PubMed Central; and
3. *Development of a web-based search and analysis engine* over the semantic provenance repository.

At present, the ProvCaRe knowledge repository is one of the largest repository of provenance metadata with more than 166 million "provenance triples" (conforming to the RDF triple structure) that was extracted from all the available 1.6 million full-text biomedical articles in the NCBI PubMed Central database.[2] All these 166 million semantic provenance triples have been mapped to the ProvCaRe provenance ontology and consists of more than 129 million unique provenance terms extracted from published articles describing a variety of biomedical and health-related research studies. Figure 5.1 shows how the provenance triples are connected to each other through common provenance terms—the total number of provenance triples is always greater than the unique provenance terms extracted from the PubMed Central.

The ProvCaRe knowledge repository is a unique resource that can enable the development of a systematic approach to evaluate and characterize the scientific reproducibility in a comprehensive manner.

---

[1] https://provcare.case.edu.

[2] http://www.ncbi.nlm.nih.gov/entrez/query.fcgi?db=PubMed.

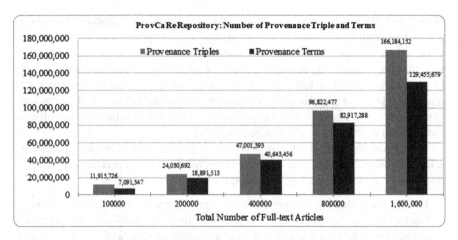

Fig. 5.1 Distribution of the provenance triples and unique provenance terms extracted from different-sized corpus of full-text articles in the ProvCaRe project

5.1.1 Related Work

The role of provenance in supporting reproducibility has been highlighted by the extensive work in scientific workflow systems, such as Taverna (Missier et al. 2010), Kepler (Ludäscher et al. 2005), and Trident (Barga et al. 2010). These workflow systems systematically record various aspects of the workflow executions, including the input/output parameters, the data used in a workflow, and pre- and post-conditions. The W3C PROV specifications, including the *PROV Data Model (PROV-DM)*,[3] *PROV-O*,[4] and the *PROV Constraints*,[5] were developed to facilitate a common approach for modeling provenance metadata on the Web. Although there have been several initiatives focused on scientific reproducibility in the biomedical and health research domains, there has been limited work in terms of the development of a general provenance metadata-based approach for scientific reproducibility.

For example, the *Ontology for Clinical Research (OCRe)* project developed an ontology to model clinical trials by incorporating various aspects of clinical research, including study design and interpretation of data (Sim et al. 2014). The *Biomedical Research Integrated Domain Group (BRIDG)* model (Fridsma et al. 2008) and the *Eligibility Criteria Extraction and Representation (EliXR)* project (Weng et al. 2011) have focused on clinical trials that identify persons meeting specific criteria for use of new drugs or therapeutic approaches. The NIH

---

[3]https://www.w3.org/TR/prov-dm/.

[4]https://www.w3.org/TR/prov-o/.

[5]https://www.w3.org/TR/prov-constraints/.

"Rigor and Reproducibility" guidelines[6] focus on important aspects of scientific research, including the source of material used in an experiment, the method used to record as well as process data and data analysis techniques. It is important to note that the components of the Rigor and Reproducibility guidelines correspond to provenance metadata terms, which validates the importance of provenance in scientific reproducibility. However, there are no provenance models that can be used to practically implement the NIH Rigor and Reproducibility guidelines in software tools or frameworks for scientific reproducibility.

In addition, the lack of a real-world provenance metadata repository describing scientific research studies also impedes the development of a practical provenance query and analysis approaches, for example:

1. Metrics for evaluating the probability of reproducing a given scientific study;
2. Characterizing the properties of a scientific research study (or best practices) reported in a published article that support reproducibility; and
3. Identifying missing provenance metadata that prevent the reproducibility of results of a research study.

5.1.2   Overview of the ProvCaRe Resource for Semantic Provenance for Scientific Reproducibility

The ProvCaRe resource was developed to address the challenges in using provenance metadata for scientific reproducibility in the biomedical and health domains. In addition to the semantic provenance repository, the ProvCaRe resource features an ontology-driven provenance query and exploration portal with a novel interactive provenance-based ranking feature. The provenance ranking feature allows users to search for research studies that have made high quality provenance information available to support reproducibility of their published results. In particular, the ProvCaRe search and query interface has been developed to allow biomedical researchers to search for previous research studies corresponding to their research hypothesis. Provenance information associated with each research study in the query result is available for review in the user interface and also for download for subsequent analysis. Figure 5.2 shows the overall provenance modeling, extraction, repository, and query interface in the ProvCaRe platform that are explained in detail in the following sections.

---

[6]https://www.nih.gov/research-training/rigor-reproducibility/principles-guidelines-reporting-preclinical-research.

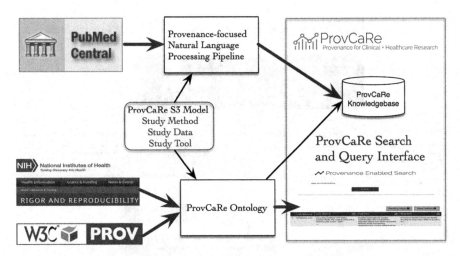

Fig. 5.2 Architecture of the semantic provenance extraction and creation of the ProvCaRe knowledge base

5.2 Development of the ProvCaRe Ontology

As we discussed in Sect. 5.1.1, the existing models for metadata information associated with biomedical and health domain have focused on clinical trials. To the best of our knowledge, there are no existing ontologies that extend the W3C PROV-O to model provenance information required to support scientific reproducibility in the biomedical and health domains. We describe the approach used in this project to develop the ProvCaRe S3 model and a corresponding provenance ontology for supporting scientific reproducibility.

5.2.1 The ProvCaRe S3 Model

The NIH Rigor and Reproducibility guidelines focus on four key aspects of scientific reproducibility: (1) soundness of prior research related to a specific hypothesis with identification of weakness of previous research studies; (2) rigor of the proposed research design with a focus on ensuring the generation of unbiased results; (3) research study data; and (4) authentication of the various resources used in a research study, including various tools used in an experiment. We generalized these key components of the Rigor and Reproducibility guidelines into the three core concepts of a provenance model for biomedical health domain, i.e., (1) Study Method, (2) Study Data, and (3) Study Tool, which together form the core components of the ProvCaRe S3 model. In addition, we used the conceptual framework of the Population, Intervention, Comparison, Outcome, and

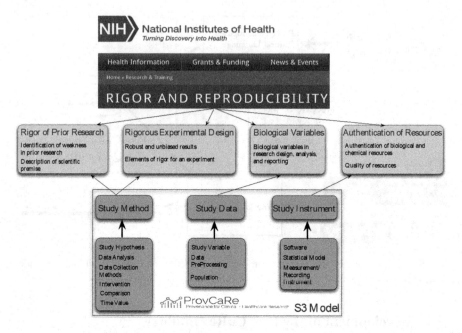

Fig. 5.3 The NIH Rigor and Reproducibility guidelines and the ProvCaRe S3 model

Time (PICO(T)) framework, which is widely used to model research questions in medical research, to expand each of the three core terms of the S3 model with various subcategories of provenance terms. We identified several limitations of the PICO(T) model that limited its effectiveness in supporting scientific reproducibility, and defined the necessary provenance terms in the S3 model to address these limitations. Figure 5.3 shows the three parent concepts and the corresponding subcategories of terms of the ProvCaRe S3 model with mappings to the NIH Rigor and Reproducibility guidelines terms.

The conceptual S3 model was formalized in the ProvCaRe ontology, which is aligned with W3C PROV-O.

5.2.2 ProvCaRe Ontology

The ProvCaRe ontology extends the W3C PROV-O with terms defined in the S3 model, which are provenance terms describing different aspects of a research experiment across multiple biomedical disciplines. For example, terms to describe the different methods used for data collection, data analysis, design of the study, and the inclusion/exclusion criteria for selecting participants in a research study

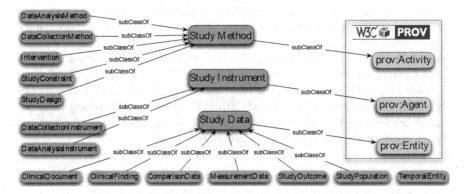

Fig. 5.4 Hierarchy of the ProvCaRe ontology consisting of parent classes that are aligned with W3C PROV-O concepts

(modeled as subclasses of the `provcare:StudyMethod` class).[7] The different instruments used for collecting and analyzing the data are modeled as subclasses of the `provcare:StudyInstrument` class. Similarly, the various categories of data associated with a research study, including the clinical findings, the outcome of the study, and comparison and measurement data are modeled as the subclasses of the `provcare:StudyData` class. A snapshot of the ProvCaRe ontology class hierarchy is shown in Fig. 5.4.

In addition to these provenance terms that are applicable across the biomedical sub-disciplines, the design pattern of the ProvCaRe ontology supports the modeling of additional provenance terms for other biomedical disciplines. For example, in collaboration with the largest repository of sleep medicine research study data in the National Sleep Research Resource (NSRR) (Dean et al. 2016), we modeled provenance terms that are essential for research studies in the sleep medicine domain. The objective of the NSRR project is to make available data collected in 40,000 research studies from more than 36,000 participants to the sleep medicine community to expedite research. Therefore, it is intuitive that the complementary availability of the provenance metadata from the research studies that have made their data available through the NSRR will facilitate reproducibility of the results published by these research studies. Similarly, the ProvCaRe ontology currently models provenance terms describing the biomedical disciplines of endocrinology and neurology. We note that we have developed a post-coordination grammar to support the extension of the ProvCaRe ontology for different disciplines of biomedicine instead of using a pre-coordinated ontology class hierarchy, which would have been both impractical and difficult to maintain in the future. We refer to our earlier publication for the details of the post-coordination grammar syntax (Valdez et al. 2017).

---

[7]The `provcare` namespace refers to `http://www.case.edu/ProvCaRe/provcare#`.

The ProvCaRe ontology uses the RDFS utility property `rdfs:seeAlso` to map its classes to classes in existing ontologies. In particular, many of the discipline-specific provenance classes are linked to classes in the SNOMED CT (Systematized Nomenclature of Medicine Clinical Terms) ontology, which is used as a de facto standard for ontology modeling in Electronic Health Record (EHR) systems (Giannangelo and Fenton 2008). The ProvCaRe ontology classes are also linked to the OCRe ontology classes (Sim et al. 2014). These links with existing ontology classes are leveraged by the ProvCaRe Natural Language Processing pipeline for named entity recognition (NER), as described in the next section.

5.3   A Provenance-Focused Ontology-Driven Natural Language Processing Pipeline

In contrast to several projects in the biomedical NLP domain that focused only on the abstract of published articles, we processed and extracted provenance information from the full text of the published articles. Table 5.1 shows the distribution of provenance terms extracted from different sections of the 1.6 million full-text articles that were processed by the ProvCaRe NLP pipeline.

Although the results show that the Method and Result sections of the papers contributed most provenance terms, a significant number of provenance terms have been extracted from the Conclusion section as well (in addition to the Abstract section that is traditionally parsed in biomedical NLP projects). The processing of full-text articles in the ProvCaRe project ensures that provenance metadata extraction from these published articles in systematic and comprehensive.

5.3.1   *ProvCaRe NLP Pipeline*

To the best of our knowledge, there are no existing biomedical NLP tools that can identify and extract provenance metadata terms from unstructured text. Therefore, we implemented a unique provenance-focused NLP pipeline by extending the open source *Apache Clinical Text Analysis Knowledge Extraction System (cTAKES)* (Savova et al. 2010), which is in turn built on the Apache Unstructured

Table 5.1 Distribution of provenance terms extracted from different sections of the full-text published articles by the ProvCaRe NLP pipeline

Paper section	Number of unique provenance terms extracted
Abstract section	1.9 million terms
Method section	5.2 million terms
Result section	4.1 million terms
Conclusion section	1.5 million terms

Information Management Architecture (UIMA). The objective of the ProvCaRe NLP pipeline is to accurately identify provenance terms corresponding to the three parent classes of the ProvCaRe S3 model, extract the provenance terms, and generate structured provenance information using an RDF-compatible triple structure. These provenance triples, for example *SleepData—wasMeasuredUsing—EpworthSleepinessScale*, correspond to the intuitive graph structure that is recommended by the W3C PROV specifications. In addition, the aggregation of these provenance triples results in the creation of a unified provenance graph that can be queried by users to search across the biomedical research domain.

The ProvCaRe NLP pipeline reuses some components of the cTAKES tool, for example, sentence boundary detector, tokenizer, and a part of the speech tagger modules, whereas we use a combined approach for provenance named entity recognition (NER). In particular, the NLP pipeline uses the ProvCaRe ontology for NER together with the NIH MetaMap tool (Aronson and Lang 2010) and the National Center for Biomedical Ontologies (NCBO) Open Biomedical Annotator (OBA) (Jonquet et al. 2009). The output of the provenance NER module in the ProvCaRe NLP pipeline is processed using a dependency parser and Semantic Role Labeler (SRL) to generate a parse tree as well as SRL-labeled predicates to form the predicate of the provenance triples. The different components of the provenance triples are mapped to the ProvCaRe ontology classes, and the provenance triples are stored in the ProvCaRe knowledge base (at present, there are 166 million provenance triples available in the ProvCaRe knowledge base). The different features of the user interface developed for the ProvCaRe knowledge base are described in the next section.

5.4 ProvCaRe Knowledge Repository and Provenance-Based Ranking

The web-accessible ProvCaRe user interface[8] features a rich set of functionalities to enable users to search and download provenance information associated with published research studies. In particular, the ProvCaRe user interface is designed to support users to conduct a "provenance-based" literature survey for their research hypothesis. This provenance-driven approach uses the provenance information extracted by the ProvCaRe NLP pipeline to rank published articles based on their "reproducibility rank." The ProvCaRe user interface features a novel provenance-based ranking dashboard that allows users to assign customized weights to different categories of provenance information (based on the ProvCaRe S3 model), which is used by the ProvCaRe ranking algorithm to display the query results to users. We describe the query and ranking features of the ProvCaRe knowledge in the following section.

---

[8]https://provcare.case.edu.

5.4.1 ProvCaRe Query Interface

The query interface for the ProvCaRe knowledge base was developed with a focus on biomedical researchers with a "free text" search interface that is similar to a web search engine interface. The query engine maps the terms in the user query to the ProvCaRe ontology classes, and uses an ontology-based query expression expansion approach to include all subclasses of the mapped term in the ontology. Using *OWLAPI*,[9] the ProvCaRe search engine also includes external ontology classes linked to the ProvCaRe ontology class by traversing graph edges having the `rdfs:seeAlso` property. Figure 5.5 shows a screenshot of the query interface with an example user query, the result of the search query, and features to save the query as well as the interactive interface for assigning weights to provenance terms in the ranking algorithm.

The results of the user queries are articles describing a relevant research study together with the provenance metadata extracted by the ProvCaRe NLP pipeline. As shown in Fig. 5.5, the provenance triples extracted from the paper are categorized into one of the three core categories of the ProvCaRe S3 model. In addition, the users can review the complete list of provenance triples corresponding to each category of the S3 model, as shown in Fig. 5.6.

Fig. 5.5 The ProvCaRe query interface with various features available to users to save queries and assign custom provenance ranks to query results

---

[9]https://github.com/owlcs/owlapi/.

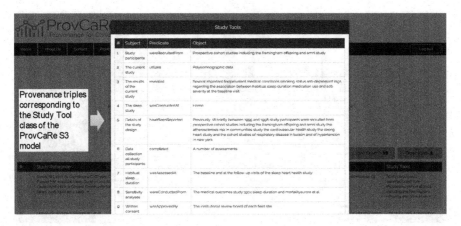

Fig. 5.6 Users can review in detail the provenance triples that have been extracted from an article and categorized according to one of the three S3 model classes

Figure 5.6 shows the provenance triples consisting of subject, predicate, and object that have been extracted from an article by Ramprakash et al. related to sleep disordered breathing. The query results of a user query typically consist of hundreds or even thousands of articles; therefore, there is a clear need to effectively rank the query results to allow users quickly locate relevant research studies. The ProvCaRe query engine uses an information retrieval-based F-measure to assign ranks to each article in the result set based on the provenance metadata extracted from an article with a default assignment of equal weights to the three core classes of the S3 model (Sahoo et al. 2019). However, if needed, users can assign customized weights to the different core classes of the S3 model using an interactive provenance ranking dashboard.

5.4.2 Provenance-Based Ranking

The ProvCaRe ranking algorithm assigns equal weight to provenance triples categorized into the three S3 model classes, i.e., a weight of 0.33 to Study Method, Study Data, and Study Instrument, in the default setting. However, users may prefer to assign more weight to the tool used to record the experiment data in comparison with the design of the research study. Similarly, some users may prefer to assign higher weight to provenance information describing the statistical analysis method used to generate the final experiment results. To support these user-defined weight assignments to the provenance-based ranking algorithm in the ProvCaRe query engine, a new provenance ranking dashboard has been developed.

Figure 5.7 shows a screenshot of the default assignment of weights to the three classes of the S3 model in the provenance ranking dashboard.

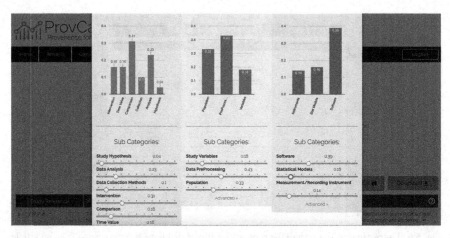

Fig. 5.7 The provenance ranking dashboard with default values assigned to the three S3 category classes

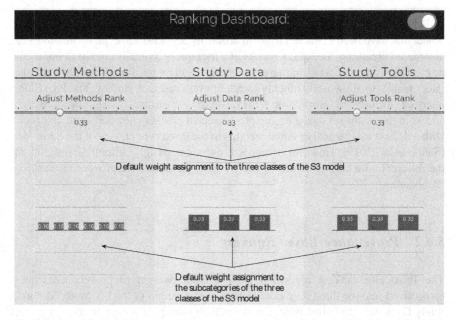

Fig. 5.8 Screenshot of different weights assigned to all the subclasses of the three S3 model classes using the provenance ranking dashboard

Figure 5.8 shows the fully expanded provenance ranking dashboard that allows users to not only assign custom weights to the 3 classes of the S3 model, but also to the subclasses of the S3 model classes.

The provenance ranking dashboard features an "auto-weight adjustment" feature that adjusts the weights of other two classes of the S3 model when weights are increased or decreased for one of the S3 classes. Users can turn off this feature to allow them to flexibly assign customized weights to the S3 classes and subclasses.

To the best of our knowledge, this is the first implementation of a provenance ranking dashboard and the incorporation of provenance weights in the ranking of user query results. At present, we are systematically characterizing the impact of user-defined provenance weights on the ranking algorithm using a set of domain expert-defined user queries.

5.5 Discussion

We have used semantic technologies to create one of the largest repositories of real-world provenance metadata extracted from all of the available 1.6 million full-text articles to support scientific reproducibility in the biomedical and health domains. The extracted provenance information is structured and represented as provenance triples, which can be aggregated to form a provenance graph.

We developed a formal model of provenance information called the S3 model, which was formalized as the ProvCaRe ontology by extending W3Cs PROV-O. Using the ProvCaRe ontology as a knowledge reference model, we developed a provenance-focused NLP pipeline that not only identifies provenance information in unstructured text, but also generates structured provenance information as provenance triples, which can be aggregated to form provenance graphs. The ProvCaRe NLP is batch-executed to periodically update the ProvCaRe knowledge repository as new full-text articles are added to the PubMed Central repository. The ProvCaRe query engine allows users to perform a provenance-enabled literature survey to identify prior research studies corresponding to their hypothesis that have a higher probability of reproducibility based on the number and category of provenance information extracted from the corresponding published article.

We are currently querying and analyzing the provenance triples in the ProvCaRe knowledge repository to identify missing provenance information in articles that are ranked low for specific user queries (we have created a "query bank" in collaboration with domain users). Similarly, we are also analyzing articles that are ranked high for user queries to identify "best practices," which can be shared with the biomedical research community to improve reporting guidelines and collection of provenance information during the conduct of scientific experiments.

5.6 Conclusion and Future Work

The growing focus on scientific reproducibility has led to a significant interest in the ProvCaRe knowledge base and the ProvCaRe ontology. As part of our ongoing work, we are developing a new summarization algorithm to concisely represent verbose phrases in the provenance triples, especially the object component of triples, which will allow us to serialize the provenance triples as RDF triples. The transformation of the ProvCaRe provenance triples into RDF triples will allow us to leverage additional semantic technologies, including SPARQL query execution and RDF graph query methods for provenance queries.

We also plan to develop a new feature in the ProvCaRe user interface to allow users to input new provenance terms to be added to the ProvCaRe ontology using the post-coordination grammar for additional biomedical disciplines. Finally, we propose to perform a systematic validation of the ProvCaRe provenance information with the available research study result in existing biomedical data repositories, such as NSRR. This evaluation will demonstrate the role of provenance information in supporting scientific reproducibility with quantitative data.

In conclusion, we have presented a semantic provenance knowledge base consisting of one of the largest repositories of provenance information extracted from unstructured text to support scientific reproducibility in the biomedical and health domains. We expect the ProvCaRe resource to serve researchers in the biomedical domain as well as the wider scientific community to evaluate and characterize scientific reproducibility of published research studies using essential contextual information captured by provenance metadata. Therefore, our expectation is that the resource contributed in this paper facilitates development of new tools for provenance-enabled scientific reproducibility.

Acknowledgments This work is supported in part by the NIH-NIBIB Big Data to Knowledge (BD2K) 1U01EB020955 grant and NSF grant #1636850.

References

(2020) The National Institute of Mental Health Data Archive (NDA). https://data-archive.nimh.nih.gov

Aronson AR, Lang FM (2010) An overview of MetaMap: historical perspective and recent advances. J Am Med Inform Assn 17(3):229–236. https://doi.org/10.1136/jamia.2009.002733

Baker M (2016) 1,500 scientists lift the lid on reproducibility. Nature 7604:452–454. https://doi.org/10.1038/533452a

Barga R, Simmhan Y, Withana EC, Sahoo S, Jackson J, Araujo N (2010) Provenance for scientific workflows towards reproducible research. IEEE Data Eng Bull 33:50–58. https://cci.case.edu/cci/images/e/e5/Barga.pdf

Begley CG, Ellis LM (2012) Raise standards for preclinical cancer research. Nature 483:531–533. https://doi.org/10.1038/483531a

Collins FS, Tabak LA (2014) Policy: NIH plans to enhance reproducibility. Nature 505(1):612–613. https://doi.org/10.1038/505612a

Dean DA, Goldberger AL, Mueller R, Kim M, Rueschman M, Mobley D, Sahoo SS, Jayapandian CP, Cui L, Morrical MG, Surovec S, Zhang GQ, Redline S (2016) Scaling up scientific discovery in sleep medicine: The national sleep research resource. Sleep 39(5):1151–1164. https://doi.org/10.5665/sleep.5774

Fridsma DB, Evans J, Hastak S, Mead CN (2008) The BRIDG Project: a technical report. J Am Med Inform Assn 5(2):130–137. https://doi.org/10.1197/jamia.M2556

Giannangelo K, Fenton SH (2008) SNOMED CT survey: an assessment of implementation in EMR/EHR applications. Perspect Health Inf Manag 5(7). https://pubmed.ncbi.nlm.nih.gov/18509501/

Jonquet C, Shah NH, Musen MA (2009) The Open Biomedical Annotator. In: Summit on Translat Bioinforma. AMIA, Bethesda, pp 56–60. https://pubmed.ncbi.nlm.nih.gov/21347171/

Kilkenny C, Browne WJ, Cuthill IC, Emerson M, Altman DG (2010) Improving bioscience research reporting: the ARRIVE guidelines for reporting animal research. PLoS Biol 8(6), e1000412. https://doi.org/10.1371/journal.pbio.1000412

Landis SC, Amara SG, Asadullah K, Austin CP, Blumenstein R, Bradley EW, Crystal RG, Darnell RB, Ferrante RJ, Fillit H, Finkelstein R, Fisher M, Gendelman HE, Golub RM, Goudreau JL, Gross RA, Gubitz AK, Hesterlee SE, Howells DW, Huguenard J, Kelner K, Koroshetz W, Krainc D, Lazic SE, Levine MS, Macleod MR, McCall JM, Moxley RT III, Narasimhan K, Noble LJ, Perrin S, Porter JD, Steward O, Unger E, Utz U, Silberberg SD (2012) A call for transparent reporting to optimize the predictive value of preclinical research. Nature 490:187–191. https://doi.org/10.1038/nature11556

Ludäscher B, Altintas I, Berkley C, Higgins D, Jaeger E, Jones M, Lee EA, Tao J, Zhao Y (2005) Scientific workflow management and the Kepler system. Concurr Comput Pract Exper 18:1039–1065. https://doi.org/10.1002/cpe.994

Missier P, Soiland-Reyes S, Owen S, Tan W, Nenadic A, Dunlop I, Williams A, Oinn T, Goble C (2010) Taverna, reloaded. In: Gertz M, Ludäscher B (eds) Scientific and statistical database management. Springer, Heidelberg, pp 471–481. https://doi.org/10.1007/978-3-642-13818-8_33

Munafò MR, Nosek BA, Bishop DVM, Button KS, Chambers CD, du Sert NP, Simonsohn U, Wagenmakers EJ, Ware JJ, Ioannidis JPA (2017) A manifesto for reproducible science. Nat Hum Behav 1(1):1–9. https://doi.org/10.1038/s41562-016-0021

Sahoo SS, Valdez J, Kim M, Rueschman M, Redline S (2019) ProvCaRe: characterizing scientific reproducibility of biomedical research studies using semantic provenance metadata. Int J Med Inform 121:10–18. https://doi.org/10.1016/j.ijmedinf.2018.10.009

Savova GK, Masanz JJ, Ogren PV, Zheng J, Sohn S, Kipper-Schuler KC, Chute CG (2010) Mayo clinical text analysis and knowledge extraction system (cTAKES): architecture, component evaluation and applications. J Am Med Inform Assoc 17(5):507–513. https://doi.org/10.1136/jamia.2009.001560

Schulz KF, Altman DG, Moher D (2010) CONSORT 2010 Statement: updated guidelines for reporting parallel group randomised trials. J Clin Epidemiol 63(8):834–840. https://doi.org/10.1016/j.jclinepi.2010.02.005

Sim I, WTu S, Carini S, PLehmann H, HPollock B, Peleg M, Wittkowskif KM (2014) The Ontology of Clinical Research (OCRe): an informatics foundation for the science of clinical research. J Biomed Inform 52:78–91. https://doi.org/10.1016/j.jbi.2013.11.002

Stodden V, Guo P, Ma Z (2013) Toward reproducible computational research: an empirical analysis of data and code policy adoption by journals. PLoS One 8(6), e67111. https://doi.org/10.1371/journal.pone.0067111

The Cancer Genome Atlas Research Network, Weinstein JN, Collisson EA, Mills GB, Shaw KRM, Ozenberger BA, Ellrott K, Shmulevich I, Sander C, Stuart JM (2013) The Cancer Genome Atlas Pan-Cancer Analysis Project. Nat Genet 45:1113–1120. https://doi.org/10.1038/ng.2764

Valdez J, Rueschman M, Kim M, Arabyarmohammadi S, Redline S, Sahoo SS (2017) An extensible ontology modeling approach using post coordinated expressions for semantic provenance in biomedical research. In: Panetto H, Debruyne C, Gaaloul W, Papazoglou M, Paschke A, Ardagna CA, Meersman R (eds) On the move to meaningful internet systems. OTM 2017 conferences. Springer, Cham, pp 337–352. https://doi.org/10.1007/978-3-319-69459-7_23

Weng C, Wu X, Luo Z, Boland MR, Theodoratos D, Johnson SB (2011) EliXR: an approach to eligibility criteria extraction and representation. J Am Med Inform Assn 18(S1):i116–i124. https://doi.org/10.1136/amiajnl-2011-000321

Chapter 6
Graph-Based Natural Language Processing for the Pharmaceutical Industry

Alexandra Dumitriu, Cliona Molony, and Chathuri Daluwatte

6.1 Introduction

The growing amount of unstructured text data available to the pharmaceutical industry in digital format has led to an increased demand of natural language processing (NLP) methods for medical evidence generation. A multitude of NLP, machine learning (ML), and artificial intelligence (AI) techniques has been used to address this need.

Knowledge graphs (KGs) represent information in the form of triplets, consisting of subject–relation–object tuples. This information can be domain knowledge (e.g., in the medical domain, Paracetamol–reduces–fever) or commonsense knowledge (e.g., Paris–capitalOf–France). While grounding information from knowledge graphs can enhance NLP application performance, it is still a challenging task, given the size and limitations of KGs in terms of completeness. Consequently, although graph-based NLP solutions have shown a lot of promise, they are still used primarily as proof of concepts, and have been considered only sparingly for the biomedical domain (Callahan et al. 2020).

Unstructured text data relevant to, and frequently analyzed by, the pharmaceutical industry is complex (see Fig. 6.1) because of being linguistically varied; i.e., the same medical event can be described differently by patients, healthcare professionals, scientists, regulatory and policy makers, and the general public.

Additionally, linguistic variation can occur within the same dataset (e.g., case report narratives used for pharmacovigilance—see section "6.5 Application Area 4: Pharmacovigilance" for more details), which makes it hard to transfer methods from one text-based dataset to another, even within the same domain. Historically,

A. Dumitriu · C. Molony · C. Daluwatte (✉)
Sanofi US Services, Inc., Cambridge, MA, USA
e-mail: Chathuri.Daluwatte@sanofi.com

© Springer Nature Switzerland AG 2021
L. F. Sikos et al. (eds.), *Provenance in Data Science*, Advanced Information
and Knowledge Processing, https://doi.org/10.1007/978-3-030-67681-0_6

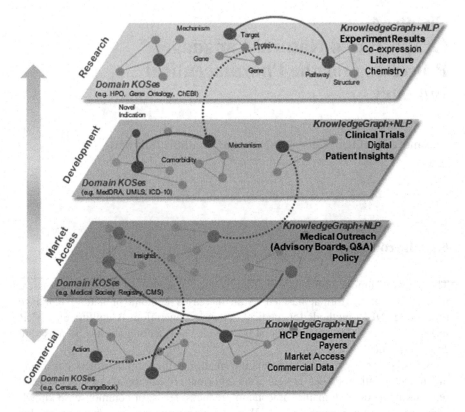

Fig. 6.1 Illustration of complex knowledge organization systems (KOSes), unstructured text data, and related structured data in pharmaceutical industry

dictionaries and ontologies (e.g., Medical Dictionary for Regulatory Activities or MedDRA), standards (e.g., International Classification of Diseases or ICD), and related compendia (e.g., Unified Medical Language System or UMLS) have been developed and maintained to address such linguistic variation. These are developed and maintained by organizations such as the World Health Organization (WHO), the National Library of Medicine (NLM), and the International Federation of Pharmaceutical Manufacturers & Associations (IFPMA) and have been endorsed by global organizations to encourage harmonization of linguistic variation (as evidenced by the International Conference on Harmonization of Technical Requirements for Registration of Pharmaceuticals for Human Use (ICH) endorsement of MedDRA). Such ontologies, standards, and related compendia provide the knowledge layer of graph-based natural language processing for text-focused pharmaceutical use cases, and we will refer to these as *knowledge organization systems (KOSes)*.

In this chapter, we present an overview of graph-based NLP techniques used in the pharmaceutical industry, focusing on four groups of problems: (1) topic identification in biomedically relevant text data, (2) patient identification, (3) clinical decision support applications, and (4) pharmacovigilance. For these use cases, several concrete examples of NLP applications that are or can be upgraded to graph-based methods are provided, while also discussing approaches to build and maintain knowledge graphs.

6.2   Application Area 1: Topic Identification

Many biomedically relevant text documents, such as peer-reviewed literature, online patient blogs or forums, medical inquiries, and patient surveys, contain rich information that pharmaceutical companies could learn from and continuously respond to. To achieve these objectives, textual data need to be regularly processed, summarized, and categorized. These tasks increase in complexity with large volumes or continuous flow of data, frequent use of domain-specific terms, and missing contextual details from short documents. To minimize time-consuming manual assessment and classification of text data, unsupervised NLP methods can be applied to identify topics and extract information-rich concepts.

Probabilistic topic models, such as *Latent Dirichlet Allocation (LDA)* (Blei et al. 2003) and *Probabilistic Latent Semantic Analysis (PLSA)* (Hofmann 2001), are classical unsupervised approaches commonly utilized for document comprehension. While implementation of these approaches is well established, one of the observed limitations is their tendency to generate topics (represented as multinomial distributions over words) that can be difficult to interpret (Hofmann 2001; Mimno et al. 2011). Meaningful and coherent topics are more likely to be obtained when topic models are augmented by related domain-specific knowledge (Andrzejewski and Zhu 2009).

Word semantics can provide the relevant knowledge to be incorporated into topic modeling. This knowledge can be captured in different ways, including via ontologies, more broadly exemplified by the *RDF (Resource Description Framework)*[1] data structure coupled with the *OWL (Web Ontology Language)*[2] vocabulary. Word embeddings, usually learnt from large corpora, represent text as vectors of real numbers that capture contextual relationships present in these corpora. For example, the vectors of two related words based on their semantic context (such as "diabetes" and "insulin") are going to be mathematically close to each other in vector space (Joulin et al. 2017; Mikolov et al. 2013; Pennington et al. 2014; Rehurek and Sojka 2010). Word embeddings have been used for many NLP tasks, including syntactic parsing and sentiment analysis (Socher et al. 2013a,b), and they have also been adopted by the topic modeling field (Cao et al. 2015; Hinton and Salakhutdinov 2009; Nguyen et al. 2015; Srivastava et al. 2013). For example, Nguyen et al. (2015) introduced the *Latent Feature LDA (LF-LDA)* method, which

---

[1]https://www.w3.org/RDF/

[2]https://www.w3.org/2001/sw/wiki/OWL

uses the *word2vec* algorithm (Mikolov et al. 2013) to extend LDA, by adding latent feature vector representations of words to the Dirichlet multinomial component that generates words from topics.

Entities and relations from knowledge graphs can also be encoded into numerical representations used for downstream applications (Bordes et al. 2013, 2011; Guo et al. 2015; Speer et al. 2017; Wang et al. 2014; Yao et al. 2017). Similar to words from text corpora, although with a higher degree of implementation variability, KGs can be mapped into continuous vectors that simplify knowledge representation, while preserving the inherent structure of the original graph. There are various embedding methods that can capture the semantics of networks (Annervaz et al. 2018). The original KG embedding approach, the *TransE* model (Bordes et al. 2013), represents relations in the graph as translations in the embedding space. The resulting vectors are obtained by minimizing a global loss function regarding all entities and relations, so that each entity vector captures both global and local structural patterns of the original KG. More specifically, if we consider graph triplets as "subject–relation–object" tuples, TransE-based vectors ensure that the equation subject + relation = object holds true. This means, for example, that if the triplets Paris–capitalOf–France and Berlin–capitalOf–Germany are present in the embedded KG, the result of the vector arithmetic France–Paris+Berlin would be close to Germany. As previously shown, KG embeddings can improve the performance of machine learning algorithms and have wide applicability, particularly for text classification tasks (Annervaz et al. 2018). These embeddings can also be used to encode prior knowledge for topic modeling, as further outlined.

In 2017, Yao et al. (2017) introduced the first graph-based topic model, called *Knowledge Graph Embedding LDA (KGE-LDA)*. The authors aimed to improve topic modeling performance by obtaining the vector representations of words from external knowledge bases such as *WordNet*[3] and the now-discontinued *FreeBase* (succeeded by *Wikidata*),[4] instead of learning them from documents. The TransE algorithm was used for obtaining embeddings. Their results showed that knowledge encoded from the considered KGs captured the semantics better than the compared methods. However, this method uses the restrictive assumption that words and graph entities have identical latent representations, which will not work well when incomplete KGs are available.

In 2019, Brambilla et al. used the approach described by Yao et al. (2017) as a benchmark for additional evaluations on how to potentially improve topic modeling with the use of knowledge graphs (Annervaz et al. 2018). The authors considered two different approaches: (1) testing different multi-relational network embedding methods (*TransE* (Bordes et al. 2013), *TransH* (Wang et al. 2014), *DistMult* (Yang et al. 2014a), *PTransE* (Lin et al. 2015a), *TransR* (Lin et al. 2015b), *HolE* (Nickel et al. 2016), and *Analogy* (Liu et al. 2017)) and (2) enriching the used KG (*WN18*, a subset of WordNet (Bordes et al. 2011)) with new edges based

---

[3]https://wordnet.princeton.edu

[4]https://www.wikidata.org

on syntactic dependency relations observed between words in sentences from the same text corpus used for topic modeling (the 20-Newgroups or 20NG dataset). The second approach, KG extension with dependency trees, has the potential to address the real-world problem of KG incompleteness, in this case, by using the text corpus itself as a source of information. From a conceptual standpoint, this approach is most likely to be successful if the KG and text corpus have high content overlap. The authors reached several conclusions from their experiments: coherence and document classification accuracy generally increase with higher topic numbers, the Analogy embedding approach with high embedding dimensions showed good performance, and KG extension improved accuracy of document classification, although not topic coherence. Running similar experiments with biomedical corpora would help better understand how these results translate to applications relevant to pharmaceutical use cases. Furthermore, if found successful for the biomedical domain, the KG extension approach could potentially refine existing biomedical knowledge graphs.

The last body of research described in this section is the recent publication by Li et al., which introduces a method called *topic modeling with knowledge graph embedding (TMKGE)* (Annervaz et al. 2018). TMKGE is a Bayesian nonparametric model based on the Hierarchical Dirichlet Process (HDP) for incorporating entity embeddings from external KGs into topic modeling. By using a multinomial distribution to model the words and a multivariate Gaussian mixture to model the entities, TMKGE is more flexible than *KGE-LDA* in sharing information between words and graph entities. As a nonparametric model, TMKGE learns the number of topics and entity mixture components directly from the data. Performance for TMKGE, as well as for additional methods (*LDA* (Blei et al. 2003), *HDP* (Teh et al. 2004), and *KGE-LDA* (Yao et al. 2017)), was assessed on three data corpora: 20NG, NIPS, and Ohsumed. WN18 was used for KG embeddings. Topic coherence (using point-wise mutual information topic coherence metric (Newman et al. 2010)) and document classification outputs showed that TMKGE significantly outperforms all the other considered methods. This is a promising result that will likely lead to additional follow-on efforts for KG-based topic modeling.

Generation of more content-rich KG embeddings and other uses of KGs in machine learning are likely to see continuing growth (Ji et al. 2020). Similarly, the expansion of using KG embedding for biomedical problems, with focus on capturing information from domain-specific graphs, is expected with the increasing availability of knowledge bases (Callahan et al. 2020; Nordon et al. 2019; Rotmensch et al. 2017). While not yet commonly utilized in the context of biomedical text, topic modeling augmented by KG embeddings has the potential to enhance topic identification results, as exemplified earlier in this section. Libraries specialized on learning and evaluation of KG embeddings for the biomedical domain, such as BioKEEN (Ali et al. 2019), together with emerging approaches to improve biomedical KG embeddings (Deng et al. 2019), will likely catalyze this adoption.

6.3   Application Area 2: Patient Identification

Building improved capabilities to identify and segment patients who fit specific disease-related profiles is one of the most frequent applications in pharmaceutical research. These profiles can take different shapes: being able to flag patients who are likely to develop a specific disease (but are either asymptomatic or do not yet show all the corresponding symptoms), identifying patients who are at risk of showing severe manifestations of a disease, and finding patients who fit the right profile for clinical trials. Many times, these disease profile assessments need to be made with incomplete and heterogeneous phenotypic data, knowing that similar disease outcomes do not always map to the same biological profiles.

Identification of patients with specific diseases can be aided by knowledge graphs derived from medical records' structured and unstructured data, which capture disease–symptom relationships (Chen et al. 2020; Finlayson et al. 2014; Sondhi et al. 2012). These relationships can be causal, associative, or weighted based on the specifics of the data. The question that gets to be addressed via the resulting knowledge graphs is "Given a patient's presenting symptoms, what is the most likely diagnosis?" It was determined that KG-based disease prediction can vary widely and is correlated to factors such as sample size and the number of co-occurring diseases, with heterogeneity of patients across age and gender also having an impact. Additionally, prediction results work best when the used knowledge graphs are created in a clinical context similar to the one where the prediction is made (Chen et al. 2020; Finlayson et al. 2014; Sondhi et al. 2012).

Identifying and classifying rare disease patients are particularly difficult subtasks for this application area. Over 350 million people in the world are affected by rare diseases, which are often chronic, progressive, and life-threatening (Colbaugh et al. 2018a). Rare diseases are also difficult to diagnose, requiring an average of seven years and eight physicians to reach a diagnosis, if the diagnosis is successfully achieved at all (Pogue et al. 2018). Patients' prolonged diagnosis journeys translate into extended periods of time without proper treatment, while also delaying related goals, such as clinical trial recruitment and disease characterization (Kempf et al. 2018). Therefore, machine learning models that facilitate identification of rare disease characteristics in population-based databases, such as Electronic Health Records, claims, or social media, are very much needed.

While related machine learning efforts have already been described in the literature (Garg et al. 2016; Gunar et al. 2018; Johnson et al. 2017), creation of rare disease patient classification models is limited in terms of training and test data because of the small number of patients with confirmed diagnoses. To overcome this limitation and improve algorithm performance, creative analytical approaches need to be utilized. Two recent approaches include (1) the use of noisy, "silver standard" labeled data, represented by patients with some evidence of positive disease status (but no actual diagnosis) combined with patient feature prioritization based on embeddings derived from PubMed articles (Colbaugh et al. 2018a,b) and (2) the use of a graphical model approach that jointly models physician and patient features together with their network relationships available from prescription and medical claims data, to predict physicians likely to treat rare disease patients (Wang and Cai 2017).

The two papers by Colbaugh and colleagues propose a Cascade Prediction Algorithm that incorporates an NLP module. They focus on addressing the rare disease patient identification problem in a widely applicable setting: Electronic Health Records (EHRs) that capture patient encounters with general practitioners (GPs). For any rare disease, this type of EHRs will be limited when it comes to finding gold standard (diagnosed) patients. Additionally, lab test results will be sparse, and physician notes limited. Acknowledging that GP EHRs are likely to only contain very few (if any) diagnosed rare disease patients, the authors proposed to include in the positive-class group the patients who show some of the rare disease characteristics, without the requirement of actual diagnosis. For the negative-class group, the authors included a mix of (1) randomly sampled patients and (2) randomly sampled patients from those diagnosed with diseases with clinical presentation similar to the rare disease of interest. This is a noisy labeled cohort, since only some of these "silver standard" patients will turn out to have the correct label assigned to them. To improve predictive model creation, patient features are pruned based on how close their skip-gram embeddings are to the rare disease of interest in external medical knowledge documents (PubMed articles in this case). Afterward, an initial ensemble of decision trees classifier is trained with the considered noisy labeled data and is applied on all unlabeled EHR patients. The last step is an iterative one, where the unlabeled data in the EHR dataset is used to refine predictions and correct some noisy labels. Similarity of patients and their current labels are used for this step (similar patients are expected to have the same label). This approach was applied to a 2.5M EHR database for lipodystrophy (LD), a rare disease with a prevalence of a few individuals per million, with no diagnosed case in the used database. The obtained results were very encouraging, with 4 "likely LD" patients and 10 "possible LD" patients in the top 20 list of predictions, based on specialist assessment (Colbaugh et al. 2018b). As a connection to what KGs could provide to the approach described here, we can assume that the NLP module used for prioritization of patient features could be improved by incorporating KG-based embeddings instead of (or together with) publication-based embeddings. This assumes that the available KG is well curated for symptoms relevant to the considered rare disease. Another relevant analytical approach is the integration of KG information directly into the rare disease classification algorithm, to help compensate for the lack of labeled training examples, as exemplified in the recent work by Li et al. (2019). Li et al. focused on the classification of diseases present in less than 0.1% of the population of either of two Chinese disease diagnosis corpora. For their research, the authors developed a text classification algorithm that can assign a disease label based on the patient's narrative description of symptoms. The algorithm represents patient documents as a combination of bag of words and bag of knowledge terms, where the knowledge term is a term shared between the patient's document and the subgraph of the used KG considered for the disease classification task. The high-level intuition of this approach is that if a rare disease is present in the knowledge graph, then the KG-based information connected to this rare disease entity can be used to focus the classifier on the important elements from the patient's record. The two utilized Chinese patient diagnosis corpora were HaoDaiFu and China Re (access to datasets not available), with 51,374 patient records categorized into 805 diseases and 86,663 patient records categorized into 44 diseases, respectively. Since no well-curated medical KG was available in Chinese (the language utilized in the HaoDaiFu and China Re datasets, which would allow for direct mapping to KG entities), the authors used a general Chinese knowledge graph, *CN-DBpedia*,[5] which is automatically constructed from Chinese counterparts of Wikipedia and contains around 17M entities and 223M relationships (Xu et al. 2017). Li et al. showed that a simple approach that incorporates an imperfect knowledge graph (in terms of coverage and accuracy) can successfully improve NLP tasks for rare disease patient classification when compared to other popular methods, such as LSTM, BOW, and up-sampling. Improvements to this approach can be considered, including the use of synonyms and word embedding methods to allow for fuzzy matching between KG entities and a patient's document, which was addressed by allowing for "surrogate entities" of unmapped diseases, which are KG entities with the highest content overlap to the considered rare disease.

Overall, given the noisy, incomplete, and (many times) general information available for rare disease patients in the databases available for model development and application, the inclusion of external knowledge is likely to have a significant impact on the performance of rare disease predictive models. Li et al. showed that even an imperfect knowledge graph can help with the classification of rare diseases and implementing solutions that include highly curated knowledge graphs focused on rare diseases (or diseases with similar symptoms as the rare disease of interest) could be successful future research avenues.

6.4   Application Area 3: Clinical Decision Support

Rich clinical databases, such as EHRs and clinical data repositories (CDRs), can be valuable resources for both patient treatment decision-making and clinical research (e.g., topics related to disease etiology, quality of care, and drug utilization). Nonetheless, summarizing both structured and unstructured information captured in these resources in a way that is easily actionable and identifying pertinent external medical knowledge for related medical questions are both complex tasks. To address these challenges, clinical decision support (CDS) applications are being developed to provide healthcare professionals with relevant information to manage patient care. Questions related to health management (e.g., likely disease diagnoses based on symptoms and labs; optimal treatment regiments; test prioritization) are all

---

[5]http://kw.fudan.edu.cn/cndbpedia/download/

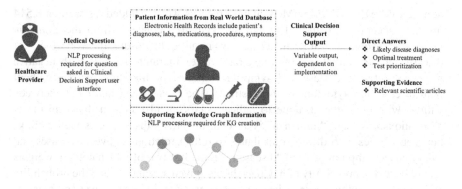

Fig. 6.2 High-level overview of knowledge graph-based clinical decision support (CDS) systems

relevant topics that CDS systems could assist with (Magrabi et al. 2019; Middleton et al. 2016; Wasylewicz and Scheepers-Hoeks 2019).

Different tools and analytical approaches have been devised to streamline the way disease-related information is extracted from EHRs, CDRs, and other rich medical databases and then synthesized and provided as guidance to the healthcare professional, clinical researcher, or even general user, as outlined in Fig. 6.2.

These tools combine NLP and KG elements and cover a range of applications, from inferring disease-related knowledge not immediately observed in the data, all the way to improving the user's mode of interaction with relevant databases (Goodwin and Harabagiu 2016; Liu et al. 2018; Ruan et al. 2019; Xia et al. 2018).

A CDS system allows for medical questions to be asked based on patient information captured in Real-World Evidence databases, such as Electronic Health Records, and supporting medical knowledge extracted and integrated into knowledge graphs from available medical resources (e.g., publications, medical databases, medical ontologies). Natural Language Processing (NLP) is needed at different steps in the process, e.g., for managing a specific query or for extracting the proper medical entities for the knowledge graph. The CDS output can assist the healthcare provider in identifying answers or supporting materials for decision-making. CDS applications can have end users beyond healthcare providers, including clinical researchers, and patients.

A first recent example, which facilitates severe disease diagnosis based on patients' symptoms, comes from Xia et al. (2018). The authors proposed an approach that utilizes a disease symptom KG constructed from the biomedical literature with the help of relationships from the *Human Disease Ontology (HDO).*[6] The approach had two steps: (1) disease–symptom relationship mapping, which consists of discovery of associations between medical entities followed by KG construction with weights for nodes and edges and (2) disease diagnosis, which uses a Bayesian inference approach with naïve independence assumptions. In total,

---

[6]https://disease-ontology.org

almost 16M MEDLINE/PubMed citation records were included, as well as 8,514 disease concepts from the HDO and 842 symptom concepts from the *Symptom Ontology*.[7] Additional filtering criteria were used to ensure the robustness of disease–symptom relationships (e.g., removal of literature records with negative mentions between disease and symptom, removal of literature records without both disease and symptom in the abstract). The created KG had edges between entities with the same attribute to indicate class membership in the form of "is a" relationships (e.g., "influenza" is a "viral infectious disease"), as well as links between entities with different attributes to reflect relations between diseases and symptoms ("symptom of" and "disease of"); the last two of which contain weights that indicate the probability of having a disease given a symptom and the probability of having a symptom given a disease (based on singleton occurrence and pair co-occurrence values from the literature data). For the second step of the approach, disease diagnosis was defined as the inference of possible diseases given available symptoms, which was mathematically computed via conditional probabilities, while assuming independence of symptoms. Given the use of literature data for knowledge graph construction, the captured information had a bias for complex or severe diseases or diseases having severe conditions. Acknowledging this limitation, Xia et al. compared the performance of their method to that of 19 other symptom checker systems on 15 standardized patient vignettes from patients known to require emergency care. The evaluation was based on whether the correct diagnosis was proposed first, proposed within the top 20 suggested diagnoses, or whether an incorrect diagnosis was made. Based on these criteria, their approach performed best. The type of symptom checker proposed by Xia et al. can be used both by patients seeking clinical information and general practitioners seeking decision support. Still, to achieve best results, the data sources utilized for KG creation need to be well suited for the considered application: patient encounter data to provide decision support for general practitioners, web-generated data for patients searching the Internet for medical information, and biomedical literature and patient encounter data of hospitalized patients for emergency or urgent care.

A different category of relevant examples falls under the topic of Question Answering (QA) using KGs (Chakraborty et al. 2019; Tong et al. 2019; Zhang et al. 2017; Zheng et al. 2018; Zheng and Zhang 2019). QA-KG is an active area of research that can facilitate the healthcare providers' interaction with relevant databases to obtain medical insights. CDS applications in this space could be responsible for the retrieval of external biomedical documents containing evidence for clinicians' questions, without going as far as extracting the answers. Alternatively, there are QA CDS applications that can provide synthesized answers to medical questions, as further outlined below.

A compelling QA CDS approach was introduced by Goodwin and Harabagiu (2016). The authors proposed a two-step framework for answering medical questions; each considered question could be represented as either a narrative describing

---

[7]https://bioportal.bioontology.org/ontologies/SYMP

fragments from the patients EMR pertinent to the medical case or a summary of the medical case. The framework first finds the most likely answer to the medical question and then selects and ranks scientific articles that contain this answer. Answer discovery is based on inference from a probabilistic knowledge graph, which was automatically generated by processing the medical language of large collections of publicly available EHRs. This NLP step identifies medical concepts representing signs/symptoms, diagnoses, tests, and treatments, together with their assertions (e.g., "present," "absent," "possible"); when this step was implemented, it created a knowledge graph of 634K nodes and 14B edges. Similar NLP steps are required to process the text of scientific articles. Goodwin and Harabagiu's approach to medical question answering produced very promising results in terms of both answer identification and relevancy of prioritized medical articles.

In 2018, Liu et al. (2018) introduced *T-Know*, a knowledge service system based on a Traditional Chinese Medicine (TCM) knowledge graph. This graph was constructed from anonymized TCM clinical records, clinical guidelines, teaching materials, medical books, and academic publications, with the use of the Bi-LSTM-CRF (Bidirectional Long Short-Term Memory with Conditional Random Field Layer) algorithm (Huang et al. 2015) for entity recognition and relation extraction. The obtained TCM KG had more than 10K nodes and 220K relations and contained five types of nodes (Diseases, Symptoms, Syndromes, Prescriptions, and Chinese Herbals). T-Know has two major modules: a QA module and a knowledge retrieval one. The user interface of the QA module allows for questions to be asked either in a single-round or multi-round (dialogue) mode. This module utilizes deep learning models to understand questions: Bi-LSTM-CRF for named entity recognition and multi-channel convolutional neural networks (CNNs) (Xu et al. 2016) for relation extraction. For the single-round QA, a joint optimization model is used to select a globally optimal "entity/relation" configuration in the TCM KG. In contrast, for the multi-round QA dialogue, the system keeps context linkage (e.g., to help with anaphora resolution). The knowledge retrieval module integrates TCM terminology and synonym dictionaries, and it allows the user to obtain relevant details about a topic of interest. For example, if the user asks about a specific disease, the module returns the disease interpretation, disease property, and other relevant information captured in the KG. Additionally, this module allows the user to visualize (in graph form) the entity-relation information available in the KG.

More recently, Ruan et al. (2019) described the *QAnalysis* tool,[8] a complex resource that allows doctors to enter analytical questions in natural language and provides back charts and tables addressing these questions. As a frame of reference, a couple of questions that can be asked (in Chinese) in the tool's user interface are "How many female patients with hyperglycemia and no hypertension have low levels of glucose tests after they took hypoglycemic drugs?" and "What is the ratio of heart failure in patients with hypertension, diabetes, and coronary heart disease?". QAnalysis consists of three modules: (1) KGs representing the available

---

[8]https://github.com/NLP-BigDataLab/QAnalysis-project

patient data and clinical terminologies, (2) a parser that transforms natural language queries into Cypher, the query language for the *Neo4j*[9] graph database management system, and (3) the Cypher queries on patient data stored in Neo4j. For the first module, the clinical data (main concepts, relations, and attributes) are captured in the OHDSI (Observational Health Data Sciences and Informatics) common data model. QAnalysis was tested on an EMR dataset available from Shanghai Shuguang Hospital, consisting of a clinical research sub-repository covering 6,035 patients with congestive heart failure over a period of 3 years; the constructed knowledge graph for this dataset had 21,728 nodes and 486,525 edges. For the second module of QAnalysis, parsing of the natural language query, several steps are required:

(a) the question entered in the user interface is segmented into bag of words, in a process that incorporates medical dictionaries;
(b) all obtained segments are annotated with different concept types (14 available concept types, including "Class," "Property," "Number," "TimeOperator," etc.);
(c) a context-free grammar is used for customized parsing to obtain dependency trees, with the Stanford parser as a fallback in case the grammar fails to parse the segments;
(d) the parse tree is transformed into a tree-like internal knowledge representation, and, if needed, a patient-schema graph is used to perform joint disambiguation; and
(e) the obtained internal knowledge representation is translated into a Cypher query statement.

The third module, represented by Cypher queries, returns statistics results—such as lists, tables, and diagrams—to the user. QAnalysis can support the most common 50 types of questions gathered from two hospital departments, and it was also shown to directly cover 165 of 211 questions (78.2%) synthesized from clinical research papers with statistical requirements for EHR data, with good quality of returned results. While there is still room for improvement with regard to the type of questions that can be successfully addressed by QAnalysis, these results are indicative of the tool's potential for simplifying statistical analysis queries in EHRs.

As summarized in this section, diverse applications of knowledge graphs can significantly contribute to improved performance of clinical decision support applications. The discussed approaches have a lot of potential, even though they need to be refined, tested on a larger scale, and implemented beyond pilots in clinical settings.

---

[9]https://neo4j.com

6.5   Application Area 4: Pharmacovigilance

Historically, post-market surveillance has been conducted largely via *individual case study reporting (ICSR)* analysis using statistical methods. However, the volume of ICSR has increased in the past decade (Stergiopoulos et al. 2019), and existing solutions still struggle to process the underlying complex knowledge of ICSR, with novel sources being considered to increase efficiency.

Individual case report narratives remain underutilized due to the underlying complexity to automate analysis of the large amount of data. Individual case reports have a rich narrative of the case reported, and the volume of reports also reveals trends and potential safety. Historically, the volume and associated trends have been analyzed, after annotating each case with the safety events it mentions. Annotating the narratives has largely been a manual task, and automatically extracting more information beyond the mentioned safety event with required accuracy remains a challenge. In addition, novel resources available for pharmacovigilance include a multitude of unstructured and structured textual data, from digital channels (Lavertu and Altman 2019) to EHR clinician notes (Liu et al. 2019). Traditional resources like pharmacovigilance textbooks, scientific literature, and drug labels are mainly explored manually. The ability of NLP methods to efficiently and automatically process these resources for pharmacovigilance evidence and hypothesis generation is currently under exploration by many (Fig. 6.3).

A multitude of NLP architectures has been used to accommodate the quality and the linguistic variations of data. All these methods try to complete three common subtasks: (1) named entity recognition (NER), (2) relation identification, and (3) relation extraction. Types of entities to recognize are adverse event, severity, indication, signs and symptoms, drug, dosage, frequency, route, and duration.

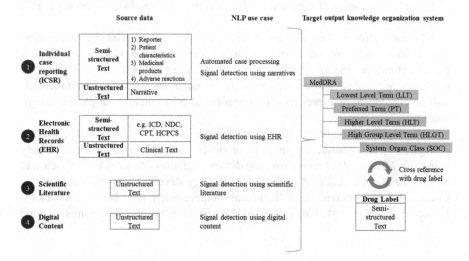

Fig. 6.3 Illustration of NLP use cases for pharmacovigilance

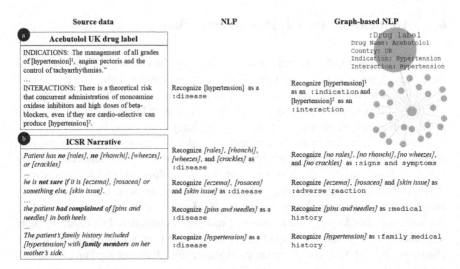

Source data	NLP	Graph-based NLP

ⓐ Acebutolol UK drug label		`:Drug label`
		`Drug Name: Acebutolol`
		`Country: UK`
		`Indication: Hypertension`
		`Interaction: Hypertension`

INDICATIONS: The management of all grades of [hypertension]¹, angina pectoris and the control of tachyarrhythmias."		
INTERACTIONS: There is a theoretical risk that concurrent administration of monoamine oxidase inhibitors and high doses of beta-blockers, even if they are cardio-selective can produce [hypertension]².	Recognize [hypertension] as a `:disease`	Recognize [hypertension]¹ as an `:indication` and [hypertension]² as an `:interaction`
ⓑ ICSR Narrative		
*Patient has **no** [rales], **no** [rhonchi], [wheezes], or [crackles]*	Recognize *[rales], [rhonchi], [wheezes],* and *[crackles]* as `:disease`	Recognize *[no rales], [no rhonchi], [no wheezes],* and *[no crackles]* as `:signs and symptoms`
*he is **not sure** if it is [eczema], [rosacea] or something else, [skin issue].*	Recognize *[eczema], [rosacea]* and *[skin issue]* as `:disease`	Recognize *[eczema], [rosacea]* and *[skin issue]* as `:adverse reaction`
*the patient **had complained** of [pins and needles] in both heels*	Recognize *[pins and needles]* as a `:disease`	Recognize *[pins and needles]* as `:medical history`
*The patient's family history included [hypertension] with **family members** on her mother's side.*	Recognize *[hypertension]* as a `:disease`	Recognize *[hypertension]* as `:family medical history`

Fig. 6.4 Illustrative examples of (**a**) semantic overlap between adverse events, indications, and signs and symptoms and (**b**) natural language constructs in pharmacovigilance that are difficult for machines to interpret

Adverse events, indications, and signs and symptoms have a semantic overlap within the dataset, since the same disease can be an indication with respect to one drug and an adverse event with respect to another. For example, in the Acebutolol UK drug label, "hypertension" is mentioned both as an indication and as an adverse event ("INDICATIONS: The management of all grades of [hypertension], angina pectoris and the control of tachyarrhythmias."; "INTERACTIONS: There is a theoretical risk that concurrent administration of monoamine oxidase inhibitors and high doses of beta-blockers, even if they are cardio-selective can produce [hypertension].") (Fig. 6.4).

NLP methodology used in pharmacovigilance can be divided into two categories: lexicon based and pattern matched (e.g., using deep learning models or distribution-based techniques) (Sarker et al. 2015). Considering the ability of a graph-based implementation of lexicon-based approach, hereafter, such lexicon-based methods will be considered as a precursor to graph-based NLP methods in pharmacovigilance.

KOSes play a significant role in pharmacovigilance. Internationally, sharing of regulatory information for medical products used by humans is done largely using the *Medical Dictionary for Regulatory Activities (MedDRA)*,[10] a rich and highly specific standardized medical terminology developed by the International Council for Harmonization of Technical Requirements for Pharmaceuticals for Human Use (ICH). MedDRA is used for the registration, documentation, and

---

[10]http://www.meddra.org

safety monitoring of medical products for both pre- and post-market surveillance; thus, safety information detected using a multitude of sources must be mapped to MedDRA. In addition to MedDRA, multiple KOSes have been used for previous graph-based NLP research for adverse event detection from unstructured text:

- *COSTART (Coding Symbols for a Thesaurus of Adverse Reaction Terms)*:[11] a vocabulary developed by the US Food and Drug Administration for post-market adverse event coding, which has been superseded by MedDRA;
- *UMLS (Unified Medical Language System)*:[12] a compendium (Metathesaurus, Semantic Network, and SPECIALIST lexicon and lexical tools) that brings together health and biomedical vocabularies and standards to enable interoperability between computer systems;
- *CHV (Consumer Health Vocabulary)*:[13] a vocabulary designed to complement the existing framework of the UMLS by connecting informal, common words and phrases about health to technical terms used by healthcare professionals;
- *ADReCS (Adverse Drug Reaction Classification System)*:[14] an adverse drug reaction compendium that provides ADR standardization and hierarchical classification (Cai et al. 2014);
- *SIDER (Side Effect Resource)*:[15] a compendium that uses MedDRA and contains adverse drug reactions extracted from public documents and package inserts; and
- *CVAD (Controlled Vocabulary-Based Adverse Drug Reaction Signal Dictionary)*: a comprehensive vocabulary developed by Lee et al. (2019), derived for vocabulary-based signal detection from EHR data, by mapping International Classification of Diseases (ICD) codes, Logical Observation Identifiers Names and Codes (LOINC), and International Classification for Nursing Practice (ICNP) mapped to the MedDRA Preferred Terms.

6.5.1   Graph-Based Natural Language Processing to Detect Adverse Events from ICSR

Post-market pharmacovigilance is conducted largely via individual case study reporting (ICSR) analysis. The first step of the process is the assessment of case validity. More specifically, a valid ICSR must meet a minimum of four criteria: (1) an identifiable reporter, (2) an individual identifiable patient, (3) at least one suspect medicinal product, and (4) at least one suspect adverse reaction. Historically, case

---

[11] https://www.nlm.nih.gov/research/umls/sourcereleasedocs/current/CST/index.html

[12] https://www.nlm.nih.gov/research/umls/index.html

[13] https://www.nlm.nih.gov/research/umls/sourcereleasedocs/current/CHV/sourcerepresentation.html

[14] http://bioinf.xmu.edu.cn/ADReCS

[15] http://sideeffects.embl.de

processing has been conducted as a demanding manual process, introducing varying quality. With the increased volume of ICSRs from the past decade (Stergiopoulos et al. 2019), Abatemarco et al. studied the ability to conduct automated case processing to increase operational efficiency and data quality consistency when processing ICSRs using an NLP and machine learning-based consortium with a business-defined threshold of F1 or accuracy $\geqslant$ 75% (Abatemarco et al. 2018). Abatemarco et al. trained ten cognitive services using NLP and machine learning: entity recognition/detection of adverse event, drug, reporter, patient and classifiers for validity, seriousness, report causality expectedness, MedDRA coding, and WHO drug dictionary (WHO-DD) coding (Abatemarco et al. 2018). Entity recognition was performed using a Bi-LSTM-CRF model, while classification was implemented using convolutional layers and a max-pooling layer. Apart from MedDRA and WHO-DD, linking to ICD was deemed useful. Such mapping can be more efficiently achieved using a graph-based NLP approach. While statistical trends run on extracted entities continue to provide valuable insights, ICSR narratives remain an underutilized resource. Extracting causality from ICSR narratives is another application for which graph-based NLP can be useful (Han et al. 2017). Apart from such conventional information, more detailed information, such as pre- and post-ADR symptoms, multi-drug usage, and adverse event relief treatments, are described in the narrative and can be extracted using graph-based NLP. Chen et al. showed that a graph-based Bi-LSTM-CRF model achieved superior performance in extracting more detailed information as entities from Chinese ICSR narratives (Chen et al. 2019).

ICSR narratives use natural language constructs that are easy to understand for humans with common sense and domain knowledge but are difficult for machines to interpret. Perera et al. described a similar problem in the context of clinical notes from EHR (Perera et al. 2013b). Linguistic and semantic components of natural language statements that should be detected in order to correctly interpret ICSRs are as follows (originally proposed by Perera et al.):

- negated statements that convey the absence of concepts, e.g., *the patient has **no** rales, **no** rhonchi, wheezes, or crackles*;
- uncertain statements that convey doubt about the presence of concept, e.g., *he/she is **not sure** if it is eczema, rosacea, or something else, skin issue.*
- statements about patient's medical history, e.g., *the patient **had complained** of pins and needles in both heels*; and
- statements about family medical history, e.g., *the patient's family history included hypertension with **family members** on her/his mother's side.*

In ICSR narratives, these mentions overlap with mentions of potential adverse events during the use of a drug. The inability of NLP engines to understand such components and associate them with relevant domain entities (e.g., no with *wheezes* and *crackles* in the first sentence) impacts the accuracy of the results, especially in the case of computer-assisted coding (Perera et al. 2013b) for ICSR case processing (Fig. 6.4). Perera et al. proposed a graph-based NLP solution (for EHR clinical notes), where "is treated with" relations are defined using UMLS, "is symptom of" relations are defined by mining EHR documents, and the knowledge base is further enriched using cTAKES. The knowledge base is expressed in the *Resource Description Framework Schema (RDFS)*[16] language. Similarly, graph-based NLP methods can be used for automated case processing in pharmacovigilance.

6.5.2   Graph-Based Natural Language Processing to Detect Adverse Events from Electronic Health Records (EHRs)

Information on post-marketing drug safety resides in electronic health records as semi-structured data coded using medical KOSes such as *ICD* or *SNOMED*,[17] as well as unstructured text as clinical notes (Sethi and Shah 2017). With the requirement to report safety information using MedDRA, medical events in EHR which are coded in ICD or SNOMED or appear in the unstructured clinical notes section must be mapped to MedDRA. Terms that describe the same medical event are in different branches of the MedDRA hierarchy, and clinically related PTs can be missed because they appear under different groups. Bousquet et al. proposed a method to knowledge engineer MedDRA vocabulary to an ontology and design definitions to enable the mapping between SNOMED CT ontology from UMLS metathesaurus as well as other medical terms that are frequently used in unstructured clinical text, therefore enabling safety surveillance definitions for MedDRA from EHR (Bousquet et al. 2019). Bousquet et al. followed NeOn Methodology (Suárez-Figueroa et al. 2012) to "ontologize" MedDRA KOS to an OWL file by converting five levels in MedDRA hierarchy into a subsumption tree and then designing definitions of MedDRA terms using several methods: (1) mapping to existing SNOMED CT from UMLS ontology, (2) mapping MedDRA terms to other sources than the UMLS (high-frequency PTs from FAERS) using semi-automatic definition algorithms, (3) decomposing MedDRA terms composed by conjunction of several distinct terms use syntactic decomposition algorithms to define MedDRA subterms in the ontology (e.g., "acute and chronic thyroiditis" is a conjunction of two medical conditions "acute thyroiditis" and "chronic thyroiditis"), (4) using lexical enrichment to generate partial definitions to MedDRA terms if a MedDRA term contains a substring with associated meaning (e.g., ending in -algia that indicate pain), and (5) providing fully manual definitions by domain experts. This process of semi-automatic and manually deriving rule-based relations based on the parents and siblings semantic definitions remains demanding and time consuming (Bousquet et al. 2019); thus, graph-based approaches can significantly increase efficiency and accuracy of this process, especially to avoid the ontology being outdated with MedDRA vocabulary updates that occur twice a year. Another approach that has been studied is the use of a common data model. The Observational Health Data

---

[16]https://www.w3.org/TR/rdf-schema/

[17]http://www.snomed.org

Sciences and Informatics (OHDSI) collaboration is developing and evaluating an informatics platform known as ADEpedia-on-OHDSI, where spontaneous reporting data from FDAs Adverse Event Reporting System (FAERS) is converted into the OHDSI CDM format (Yu et al. 2019). This project developed an extraction, transformation, and loading (ETL) tool to convert data into the OHDSI CDM format. OHDSI CDM preferred vocabulary is SNOMED CT, and under OHDSI, entities in EHR are coded into SNOMED CT. ADEpedia-on-OHDSI was first tested using data from FDA Adverse Event Reporting System (FAERS). ADEpedia-on-OHDSI failed to completely map to SNOMED CT concepts, the indications and adverse events recorded in MedDRA Preferred Terms (PTs) in FAERS. Yu et al. discussed that graph-based NLP methods can significantly improve the concept mapping between vocabularies when developing common data models for pharmacovigilance (Yu et al. 2019).

In their work, Lee et al. summarized widely used KOSes for adverse event detection from EHR. Since clinical text is written by healthcare professionals, KOSes created by healthcare professionals provide efficient results in this context (Lee et al. 2019). In addition to UMLS, MedDRA, ADReCS, SIDER, and COSTART, the following have also been used:

- *ICD (International Classification of Diseases):*[18] the international "standard diagnostic tool for epidemiology, health management, and clinical purposes," which is maintained by the WHO;
- *RxNorm:*[19] a US-specific terminology in medicine that contains all medications available on the US market. RxNorm is part of the Unified Medical Language System terminology and is maintained by the United States National Library of Medicine;
- *ATC (The Anatomical Therapeutic Chemical Classification System):*[20] a drug classification system that classifies the active ingredients of drugs according to the organ or system on which they act and their therapeutic, pharmacological, and chemical properties. It is controlled by the World Health Organization Collaborating Centre for Drug Statistics Methodology (WHOCC);
- *SNOMED Clinical Terms (SNOMED CT):* a systematically organized computer-processable collection of medical terms providing codes, terms, synonyms, and definitions used in clinical documentation and reporting. SNOMED CT is considered the most comprehensive multilingual clinical healthcare terminology in the world;
- *CPT (Current Procedural Terminology):*[21] a medical code set maintained by the American Medical Association that describes medical, surgical, and diagnostic services and is designed to communicate uniform information about medical

---

[18]https://www.who.int/classifications/icd/en/

[19]https://www.nlm.nih.gov/research/umls/rxnorm/index.html

[20]https://www.who.int/classifications/atcddd/en/

[21]https://www.ama-assn.org/amaone/cpt-current-procedural-terminology

services and procedures among physicians, coders, patients, accreditation organizations, and payers for administrative, financial, and analytical purposes;

- *HCPCS (Healthcare Common Procedure Coding System):*[22] a set of healthcare procedure codes based on the American Medical Associations Current Procedural Terminology (CPT);
- *LOINC (Logical Observation Identifiers Names and Codes):*[23] a database and universal standard for identifying medical laboratory observations, maintained by the US nonprofit medical research organization Regenstrief Institute;
- *National Procedure Classification*: national modifications of the ICD and national procedure classification schemes in the Nordic countries;
- *WHOART (WHO Adverse Reactions Terminology)*: an adverse reaction dictionary (no longer actively maintained); and
- *STITCH (Search Tool for Interactions of Chemicals):*[24] an interaction database for small molecules and proteins.

With these KOSes, the following NLP systems that consist of heuristic approaches have been used to detect adverse events from clinical text (shown in Table 6.1):

- *MedXN:*[25] it consists of fast string-matching algorithm, regular expressions, and conversion to RxNorm (Sohn et al. 2014);
- *cTAKES:*[26] it consists of sentence boundary detector, tokenizer, normalizer, part-of-speech (POS) tagger, shallow parser, and named entity recognition (NER) annotator, including status and negation annotators (Savova et al. 2010);
- *MTERMS*: it consists of exact string match, morphological analysis (e.g., handling punctuation and other morphological variations), lexical analysis (e.g., handling abbreviations and acronyms), syntactic analysis (e.g., phrase segmentation and recombination), and semantic analysis (e.g., identifying meaning and assigning the terms to an appropriate semantic group) (Zhou et al. 2011); and
- *MetaMap:*[27] tokenization, sentence boundary determination, and acronym/abbreviation identification; part-of-speech tagging; lexical lookup of input words in the SPECIALIST lexicon from UMLS; and a final syntactic analysis consisting of a shallow parse in which phrases and their lexical heads are identified by the SPECIALIST minimal commitment parser. Each phrase found by this analysis is further analyzed for (1) variant generation, in which variants of all phrase words are determined (by table lookup), (2) candidate identification, in which intermediate results consisting of metathesaurus strings, called candidates, matching some phrase text are performed and evaluated as to how well they

---

[22] https://www.cms.gov/Medicare/Coding/MedHCPCSGenInfo
[23] https://loinc.org
[24] https://www.hsls.pitt.edu/obrc/index.php?page=URL1208441111
[25] https://github.com/OHNLP/MedXN
[26] https://ctakes.apache.org
[27] https://metamap.nlm.nih.gov

match the input text, (3) mapping construction, in which candidates found in the previous step are combined and evaluated to produce a final result that best matches the phrase text, and (4) word sense disambiguation (WSD), in which mappings involving concepts that are semantically consistent with surrounding text are favored (Aronson and Lang 2010).

An additional set of NLP methods used with the mentioned KOSes utilized machine learning methods (see Table 6.1):

- *MedTagger*:[28] it utilizes UMLS and BioThesaurus as terminology resources and three machine learning models, CRF, maximum-entropy Markov model, and hidden Markov model (Yang et al. 2019);
- *MADEx*: it utilizes an LSTM-CRF model for named entity recognition with bidirectional LSTM, character-level embedding, and dropout. For relation extraction, two classifiers trained using SVMs (one for single-sentence relations and another for cross-sentence relations) were used (Yang et al. 2019).

LePendu et al. proposed a lexicon-based NLP pipeline for drug safety surveillance (both single-drug adverse events and drug–drug interactions detection) using clinical notes transformed into a feature matrix encoded using medical terminologies (LePendu et al. 2013). They used lexicons to support multi-purpose NLP of clinical notes and identify clinically relevant nouns, syntactic and semantic triggers, drugs, adverse events, drug–drug interactions, and patient cohorts. Wang et al. proposed a method that consists of a lexicon-based NLP pipeline for adverse event detection and validation and hypothesis generation (Wang et al. 2018). Lexicons based on RxNorm were used with NLP systems MedXN and MedTagger. Torii et al. used NLP programs to extract adverse events from clinical notes, followed by case-crossover design for signal detection (Torii et al. 2011). Following signal detection, the SIDER and ADReCS ontologies were used for signal validation and hypothesis generation. In their graph-based NLP pipeline, Ngo et al. used a drug dictionary created using DrugBank and SNOMED, along with preprocessing (tokenization, lemmatization, part-of-speech (POS) tagging, chunking, and dependency parsing), rule-based feature creation, and a CRF model (Ngo et al. 2018).

Deep learning neural networks have also been used to jointly model entity recognition and relation extraction (Dandala et al. 2019; Jagannatha et al. 2019; Li et al. 2018). Li et al. used deep learning (Bi-LSTM-CRF model for named entity recognition and Bi-LSTM-Attention model for relation extraction) and trained the models for multiple tasks with multi-task learning (hard parameter sharing and soft parameter sharing—task relation learning and regularization) (Li et al. 2018). The model with hard parameter sharing showed the most improvement. Dandala et al. used a Bi-LSTM-CRF model to detect entities: medication name and its attributes (dosage, frequency, route, duration), adverse events, indications, signs and symptoms, severity and a Bi-LSTM-Attention model to detect relations between these

---

[28]https://github.com/medtagger/MedTagger

entities (Dandala et al. 2019). Dandala et al. showed that jointly modeling concepts and relation extraction perform better than sequentially extracting medical entities and then relations. They further demonstrated that graph-based joint modeling approach further improved performance. For the knowledge-based joint modeling approach, Dandala et al. used unified medical language system (UMLS) concept unique identifiers (CUIs) for drugs and signs, symptoms, severity, indications, and adverse events and proportional reporting ratio and reporting odds ratio from FAERS to create the graph-based knowledge layer. In the same dataset, the graph-based Bi-LST-CRF model of Dandala et al. showed superior performance than the model of Li et al. with multi-task learning (Dandala et al. 2019; Li et al. 2018).

6.5.3   Graph-Based Natural Language Processing to Detect Adverse Events from Scientific Literature

Yeleswarapu et al. used an NLP pipeline to detect drug-adverse event pairs from Medline abstracts (Yeleswarapu et al. 2014). They used a drug dictionary and an event dictionary, based on the TPX drug dictionary, but further enhanced using Wikipedia data, MeSH, and MedDRA. This task can be efficiently achieved using graph-based NLP.

6.5.4   Graph-Based Natural Language Processing to Detect Adverse Events from Digital Content

Digital content can serve as a key resource for gathering real-world evidence on drug safety (Caster et al. 2018; Colilla et al. 2017; Gavrielov-Yusim et al. 2019; Kürzinger et al. 2018; Pierce et al. 2017; Sarker et al. 2015). The U.S. Food and Drug Administration (FDA) has recognized the utility of social media in proactive pharmacovigilance (FDA 2018; Throckmorton et al. 2018). Digital content-based pharmacovigilance has been identified to be advantageous from multiple aspects (van Stekelenborg et al. 2019):

1. It can detect rare events that are not surfaced through spontaneous reporting, alleviating underreporting in spontaneous systems. There is a potential for use as a primary tool for safety signal detection in specific areas, such as exposure during pregnancy, misuse, and abuse.
2. It can provide early detection of events.
3. It can be used to strengthen emerging hypotheses from other systems during the signal validation process.
4. It can detect events that are not medically serious/considered benign (hence, underreported in case reporting) but have an impact on quality of life, such as vitreous opacities (eye floaters).

Data from social media and web content are challenging to analyze due to the difficulty to search through large volumes of irrelevant data, high frequency of new data generation, lack of validation and user bias, and quality. Some of these aspects can be addressed with data provenance; however, it is critical that data privacy is respected (Pappa and Stergioulas 2019).

Detecting adverse events from social media and web content data provides a unique challenge due to the wide linguistic variability and quality. Previous research in graph-based NLP for adverse event detection from social media have mainly used the MedDRA, COSTART, UMLS, ADReCS, and SIDER KOSes (see Table 6.1). While there are many KOSes for pharmacovigilance, most of these have been created by healthcare professionals. Since laypersons (consumers) and professionals may use different terms to describe the same medical concept (e.g., "throwing up" vs. "vomiting"), using above mentioned KOSes with social media text data may create false negatives. Graph-based NLP can efficiently create relationships to surface layman's terms for professional healthcare terms. Some studies used the Consumer Health Vocabulary (CHV) to address this concern (Yang et al. 2012). Recent work has enhanced the CHV using GloVe embeddings generated from consumer-generated text (Ibrahim et al. 2020).

In terms of NLP methodologies, the studies can be divided into two categories: studies that used machine learning methods (NER, hidden Markov models, and classifiers) and studies that did not use machine learning methods. While NER-based studies used datasets without annotations, other studies used small datasets with annotations. Studies that did not use machine learning methods matched to lexicons using distributional-based NLP methods or rule-based methods. Most of these studies used large datasets without annotations. The focus to identify adverse events on a pure graph-based matching does not address some challenges. When consumers use creative phrases instead of technical terms, these mentions can go undetected (e.g., "messed up my sleep" instead of "sleep disturbance"). Instances where there is a match with a lexicon term, but not describing an adverse event, can create false positives. Studies such as Nikfarjam et al. (2019), Yeleswarapu et al. (2014) used entity recognition models to first detect adverse events in social media and then used graph-based lexicons to map recognized entities to vocabularies.

6.6   Knowledge Graph Development and Learning

The knowledge graphs built to support graph-based NLP with input from experts may suffer from shortcomings such as inefficiency in capturing expert knowledge, subjectivity, and incompleteness, due to knowledge gaps between knowledge engineers and domain experts (Perera et al. 2012, 2013a). Constructing knowledge graphs is a tedious process and requires demanding manual curation efforts. When the vocabularies and dictionaries used are updated frequently, repeating the graph building process manually is not feasible and requires tedious change management processes. Algorithmic (NLP and data-driven) approaches have been proposed to

Table 6.1 Summary of KOS-based NLP pharmacovigilance research

Study	Vocabulary/Ontology	NLP method
Ngo et al. (2018)	Drug dictionary using DrugBank, SNOMED	Preprocessing: tokenization, lemmatization, part-of-speech (POS) tagging, chunking, and dependency parsing Rule-based feature extraction and CRF model
Wang et al. (2018)	RxNorm, UMLS, SIDER, ADReCS	Fast string-matching algorithm, regular expressions, conversion to RxNorm using MedXN and machine learning-based detection using MedTagger
LePendu et al. (2013)	UMLS, BioPortal, trigger terms from NegEx and ConText, Nouns lexicon using MEDLINE abstracts, Medi-Span Drug Indications Database, and the National Drug File—Reference Terminology, RxNorm, EU-ADR project's specifications, and MedDRA	Identify nouns using lexicons, frequency analysis to remove uninformative phrases, syntactic and semantic suppression rules, normalize concepts using parent–child relationships
Leaman et al. (2010)	UMLS, SIDER, MedEffect, DailyStrength	Sentence breaker, tokenization, tag part of speech, remove stop words, Jaro–Winkler measurement similarity for spelling comparison
Nikfarjam and Gonzalez (2011)	COSTART	Part-of-speech tagging, sequence generation, rule-based classification
Benton et al. (2011)	Cerner Multum's Drug Lexicon, Medicinenet.com, FAERS, hand-curated dietary supplements, CHV	Natural language toolkit associated rules
Hadzi-Puric and Grmusa (2012)	EMEA, ALIMS, DrugBank, and UMLS	Concept identification
Yang et al. (2012)	CHV	Remove punctuation/stop words, tokenize words, multi-gram term generator, match with lexicons
Jiang and Zheng (2013)	MetaMAP	Machine learning classifier
Yates and Goharian (2013)	Subset of UMLS, SIDER2, MEDSyn	Match with lexicons using rule-based approach
Yeleswarapu et al. (2014)	Drug–disease–symptom dictionaries, PV-TPX, MedDRA, MeSH, Wiki	Dictionary-based NER, POS tagging, stemming, tokenizing, acronym handling, entity association
Freifeld et al. (2014)	MedDRA	Tree-based dictionary-matching algorithm

(continued)

Table 6.1 (continued)

Segura-Bedmar I (2014)	Drug dictionary created using topic extraction, AE dictionary using MedDRA and Vademecum, Drug and AE dictionary created using CIMA online information center (maintained by the Spanish Agency for Medicines and Health Products [AEMPS]), ATC as drug family dictionary	Entity recognition provided by Textalytics, followed by dictionary-based approach to identify entities, a plugin that integrates Textalytics with GATE, a general architecture for text engineering
O'Connor et al. (2014)	COSTART, FDA post-market surveillance, SIDER/SIDER 2, MedEffect, CHV	Index and retrieve concepts, tokenize, spelling, stop word removal, lemmatize
Yang et al. (2014b)	CHV, Wiki	Rule-based matching
Sampathkumar et al. (2014)	I2b2 clinical NLP, BioCreative challenges, SIDER	Named entity recognition, relation extraction, hidden Markov model

either augment graph building efforts or fully automate them (Callahan et al. 2020). In order to account for knowledge gaps, multiple sources of information have been incorporated when creating healthcare knowledge graphs. Some data-driven approaches simply transform existing databases into a knowledge graph, facilitating adherence to FAIR (Findable, Accessible, Interoperable, and Reusable). Yuan et al. proposed an approach with minimum supervision based on unstructured biomedical domain-specific contexts using entity recognition, unsupervised entity and relation embedding, latent relation generation via clustering, and relation refinement and relation assignment to assign cluster-level labels (Yuan et al. 2020). Fauqueur et al. proposed NLP methods to extract biomedical facts from the literature by leveraging and refining specific seed patterns (Fauqueur et al. 2019). Li et al. proposed a systematic NLP-based approach to construct the medical KG from EHRs (Li et al. 2020). In contrast to the classical triplet structure, Li et al. used a quadruplet structure defined as triplet of two entities and their relation. To build the graph, NER, relation extraction, and property calculation for entities and relations (occurrence number and co-occurrence number) were used with graph cleaning methods (entity and/or quadruplet deletion if properties are under prespecified thresholds). To rank and retrieve the most related entities for a given subject, subject-related entity rankings were calculated using term frequency–inverse document frequency (TF-IDF). Embedding vectors from the constructed quadruplet-based medical KG were then developed to use in clinical decision support applications. Cong et al. presented an evaluation of one of the most widely used NLP-constructed KGs, SemMedDB, and reported many contradictory assertions in a variety of fundamental relationship categories, underscoring the need to validate NLP-derived knowledge graphs (Cong et al. 2018).

Perera et al. proposed a systematic method of identifying missing relationships between concepts in knowledge graphs and generating suggestions to fill those gaps (Perera et al. 2012, 2013a). They proposed:

- a knowledge graph validation methodology by using real-world data sources,
- a bootstrapping/semi-supervised method for finding missing relationships in a knowledge graph, and
- a convenient method to elicit missing domain knowledge from experts.

The method can be summarized as follows:

1. Build initial knowledge graph. Perera et al. used *IntellegO* for this purpose. *IntellegO* is an ontology that provides formal semantics of machine perception by defining the informational processes involved in translating observations into abstractions. The ontology is encoded using the set theory. The proposed methodology used a subset of concepts (`intellego:entity`, `intellego:quality`, `intellego:percept`, and `intellego:explanation`) and the `intellego:inheres-in` relationship from *IntellegO* ontology. When building the knowledge graph, Perera et al. used `intellego:entity` and `intellego:quality` to annotate disorder and symptom, respectively.
2. Semantically annotate the EHR documents with concepts from the knowledge graph. Perera et al. used MedLEE to identify disorders, symptoms, medications, and procedures from EHR and convert these terms to XML elements. Disorders were annotated as `intellego:explanation` and symptoms as `intellego:percept`.
3. Generate the `intellego:coverage` for the document. Coverage was described as the aggregation of `intellego:quality` that can be accounted for by `intellego:explanation`.
4. Identify inconsistent EMR documents. Compare the set of `intellego:coverage` with the set of `intellego:percept` in order to discover discrepancies between the knowledge graph and EHR.
5. Suggest candidate relationships that can rectify these inconsistencies. Missing relationships are identified as `intellego:entity` and `intellego:quality`, with a missing `intellego:inheres-in`. These candidates are ranked based on how many times they co-occur with a missing `intellego:inheres-in` in inconsistent EMR documents.
6. Provide questions to domain experts about the correctness of the candidate relationships; formed as "Does A inheres-in B?" using the suggested candidate relationships.
7. Update the KG based on domain expert's feedback.

Perera et al. defined link accuracy as the "number of corrected links*100/ number of proposed links" to measure the effectiveness of their proposed method (Perera et al. 2012, 2013a).

Knowledge graph development, maintenance, and validation become further challenging for pharmaceutical industry use cases such as pharmacovigilance. The knowledge that is foundational is in semi-structured or unstructured text (e.g., drug labels) and is changing frequently (e.g. drug labels can change daily and MedDRA updates every six months). In order to be compliant, the knowledge graph always needs to be up to date with traceable changes. Data provenance can help achieve this.

6.7 Retrofitting Distributional Embeddings with Relations from Knowledge Graphs

Distributed representations of words (GloVe, ElMo, and BERT (Devlin et al. 2018; Pennington et al. 2014; Peters et al. 2018)) provide vector representations of words that help learning algorithms achieve better performance in natural language processing tasks by grouping similar words. While the distributed representation of words is very powerful, it only provides a blurred picture of the relationships between concepts. Knowledge graphs directly encode this relational information and combine the advantages of distributional and relational data. Faruqui et al. proposed to retrofit embeddings learned from distributional data to the structure of a knowledge graph (Faruqui et al. 2015). Their method first learns entity representations based solely on distributional data and then applies a retrofitting step to update the representations based on the structure of a knowledge graph. Lengerich et al. proposed "Functional Retrofitting," a framework that models pairwise relations as functions that retrofit existing distributed representations of words and concepts using the relational information in complex knowledge graphs (Lengerich et al. 2017). Lengerich et al. demonstrated their framework using pharmaceutical use cases, in knowledge accumulation and discovery, to identify new disease targets for existing drugs. Tao et al. further demonstrated how these methods can yield superior performance on pharmaceutical use cases to detect complex, multi-word healthcare concepts (Tao et al. 2018).

6.8 Graph-Based Natural Language Methods from Outside the Pharmaceutical Industry

The pharmaceutical industry has both a rich data and a question landscape to apply the techniques described above. Examples have been drawn from along the pharmaceutical value chain—research, development, medical, and operational areas. Related graph-based natural language techniques used for industries outside of pharma can be looked at for additional inspiration. Multinational companies or companies needing to work across jurisdictions will have a multitude of compliance and regulatory and legal systems to operate in simultaneously. Most companies are not scaled to operate across each of these domains with local expertise or footprint. The European innovation project Lynx aims to generate a multilingual Legal Knowledge Graph (LKG) from structured and unstructured data sources (Schneider et al. 2020). This will be combined with a set of flexible language processing services to analyze and process the data and documents to integrate them into the LKG. Existing pilots include Contracts, Labor Law, and the Oil & Gas Energy Sector. A reasonable extension is knowledge management of pharmaceutical Regulatory and Health Technology Assessment agency material. Having a broader view can enable companies to anticipate trends and identify outliers, as well as streamline internal processes.

Financial companies have embraced Natural Language Generation (NLG) to reason on data that may be stored in tabular formats and turn this information into digestible and customized reports. A typical question might be "How did Hurricane Joe impact performance of company B in the last quarter of 2019?" An AI-generated consumer or investor's report cannot simply display such a table of numbers, but it rather needs to go one step further and explain in crisp natural language the key message that addresses the user's question (Matsunaga et al. 2019; Mishra et al. 2019). Clinical Study Reports (CSRs) could be authored based on tabular-dense research results via NLG techniques combined with domain-specific KG (graph to text). Other relevant examples include autogeneration of customized content for on-demand responses for medical or commercial field and multi-channel engagement functions.

6.9 Conclusion

A variety of graph-based NLP solutions have been proposed and studied, covering several pharmaceutical industry use cases. Importantly, these methods differ widely in terms of architecture, KG development, and interfacing between KG and NLP approaches. There is no method that will work equally well on all pharmaceutical industry use cases—different methodologies and applications from various domains need to be considered when tackling biomedical text data. Although graph-based NLP solutions seem promising in different domains, they are still primarily used as proofs of concept in the pharmaceutical industry and need more research to reach the required level of maturity (in terms of overall performance levels) for real-world use.

Biomedical knowledge is extensive and can be condition-specific, and comprehensive biomedical knowledge graphs can contain billions of edges, becoming computationally expensive for regular use. Additionally, there are challenges in constructing domain-specific knowledge graphs. Some of these challenges are data availability and data licensing issues (e.g., different license for each data source), lack of de facto standards and consensus for knowledge graph construction, dependency upon shared computational resources, and specialized user's skill set or area of expertise.

The construction of knowledge graphs is a difficult, time-consuming task; therefore, reuse is highly desirable. However, unless carefully designed, there are limitations in applying an existing knowledge graph to a new task. General tools that allow for interaction between knowledge graphs can improve graph reuse in new tasks, and these tools would benefit from a central reference to identify all available biomedical knowledge graphs. Currently, except for the review by Callahan et al. (2020), the literature does not mention such an initiative. Therefore, a more systematic approach to share, document, and encourage interoperability among biomedical knowledge graphs would likely facilitate more adoption. In addition, knowledge graphs are challenging to validate, which needs to be considered because accuracy is crucial for pharmaceutical use cases. Some validation methods that have been used so far are qualitative (e.g., case studies or domain expert review of results, conceptual models or prototypes, and focus groups), quantitative (e.g., most often by applying a machine learning model to a dataset not used during model building or by performing a KG completion task like edge prediction), or a mixture of the two. Considering the frequent updates to KOSes that are critical for pharmaceutical industry use cases (e.g., MedDRA updates twice a year and drug labels change daily), using programmatic knowledge graph development, and validation methods and data provenance are extremely important for reliable real-world implementation of graph-based NLP methods.

One of the major technical challenges of using digital content in pharmaceutical industry use cases is the uncertainty and quality of data. While incorporating data provenance can help improve these aspects, with personal data protection concerns, a balancing act needs to be performed to address the respective interests of patient data protection and evidence generation.

Because of these challenges, graph-based NLP solutions for pharmaceutical use cases are still in an early stage of adoption. Nevertheless, as discussed, graph-based NLP is already utilized in some pharmaceutical use cases, and NLP methods can be used to generate and validate knowledge graphs from unstructured or semi-structured data. Applications include entity recognition, unsupervised entity and relation embedding, latent relation generation via clustering, relation refinement and relation assignment to assign cluster-level labels, extraction of biomedical facts from the literature by leveraging and refining specific seed patterns, identification of missing relationships between knowledge graph concepts, and generation of suggestions to fill those gaps.

References

Abatemarco D, Perera S, Bao SH, Desai S, Assuncao B, Tetarenko N, Danysz K, Mockute R, Widdowson M, Fornarotto N, Beauchamp S, Cicirello S, Mingle E (2018) Training augmented intelligent capabilities for pharmacovigilance: Applying deep learning approaches to individual case safety report processing. Pharm Med 32(6):391–401. https://doi.org/10.1007/s40290-018-0251-9

Ali M, Hoyt CT, Domingo-Fernández D, Lehmann J, Jabeen HJB (2019) BioKEEN: A library for learning and evaluating biological knowledge graph embeddings. Bioinform 35(18):3538–3540. https://doi.org/10.1093/bioinformatics/btz117

Andrzejewski D, Zhu X (2009) Latent Dirichlet Allocation with topic-in-set knowledge. In: Proceedings of the NAACL HLT 2009 workshop on semi-supervised learning for natural language processing, Association for Computational Linguistics, pp 43–48. https://doi.org/10.3115/1621829.1621835

Annervaz K, Chowdhury SBR, Dukkipati A (2018) Learning beyond datasets: Knowledge graph augmented neural networks for natural language processing. https://arxiv.org/pdf/1802.05930.pdf

Aronson AR, Lang FM (2010) An overview of MetaMap: Historical perspective and recent advances. J Am Med Inform Assn 17(3):229–236. https://doi.org/10.1136/jamia.2009.002733

Benton A, Ungar L, Hill S, Hennessy S, Mao J, Chung A, Leonard CE, Holmes JH (2011) Identifying potential adverse effects using the Web: A new approach to medical hypothesis generation. J Biomed Inform 44(6):989–996. https://doi.org/10.1016/j.jbi.2011.07.005

Blei DM, Ng AY, Jordan MI (2003) Latent Dirichlet allocation. J Mach Learn Res 3:993–1022. https://dl.acm.org/doi/10.5555/944919.944937

Bordes A, Weston J, Collobert R, Bengio Y (2011) Learning structured embeddings of knowledge bases. In: Twenty-fifth AAAI conference on artificial intelligence, pp 301–306

Bordes A, Usunier N, Garcia-Duran A, Weston J, Yakhnenko O (2013) Translating embeddings for modeling multi-relational data. In: Advances in neural information processing systems, pp 2787–2795. http://papers.nips.cc/paper/5071-translating-embeddings-for-modeling-multi-relational-data

Bousquet C, Souvignet J, Sadou r, Jaulent MC, Declerck G (2019) Ontological and non-ontological resources for associating medical dictionary for regulatory activities terms to SNOMED clinical terms with semantic properties. Front Pharmacol 10:975–975. https://doi.org/10.3389/fphar.2019.00975

Cai MC, Xu Q, Pan YJ, Pan W, Ji N, Li YB, Jin HJ, Liu K, Ji ZL (2014) ADReCS: An ontology database for aiding standardization and hierarchical classification of adverse drug reaction terms. Nucleic Acids Res 43(D1):D907–D913. https://doi.org/10.1093/nar/gku1066

Callahan TJ, Tripodi IJ, Pielke-Lombardo H, Hunter LE (2020) Knowledge-based biomedical data science. Ann Rev Biomed Data Sci 3:23–41. https://doi.org/10.1146/annurev-biodatasci-010820-091627

Cao Z, Li S, Liu Y, Li W, Ji H (2015) A novel neural topic model and its supervised extension. In: Twenty-ninth AAAI conference on artificial intelligence, pp 2210–2216. https://dl.acm.org/doi/abs/10.5555/2886521.2886628

Caster O, Dietrich J, Kürzinger ML, Lerch M, Maskell S, Norén GN, Tcherny-Lessenot S, Vroman B, Wisniewski A, van Stekelenborg J (2018) Assessment of the utility of social media for broad-ranging statistical signal detection in pharmacovigilance: Results from the WEB-RADR project. Drug Safety 41(12):1355–1369. https://doi.org/10.1007/s40264-018-0699-2

Chakraborty N, Lukovnikov D, Maheshwari G, Trivedi P, Lehmann J, Fischer A (2019) Introduction to neural network based approaches for question answering over knowledge graphs. https://arxiv.org/pdf/1907.09361.pdf

Chen Y, Zhou C, Li T, Wu H, Zhao X, Ye K, Liao J (2019) Named entity recognition from Chinese adverse drug event reports with lexical feature based BiLSTM-CRF and tri-training. J Biomed Inform 96:103,252. https://doi.org/10.1016/j.jbi.2019.103252. http://www.sciencedirect.com/science/article/pii/S1532046419301716

Chen IY, Agrawal M, Horng S, Sontag D (2020) Robustly extracting medical knowledge from EHRs: A case study of learning a health knowledge graph. In: Pacific symposium on biocomputing 2020. World Scientific, pp 19–30. https://doi.org/10.1142/9789811215636_0003

Colbaugh R, Glass K, Rudolf C, Global MTV (2018a) Learning to identify rare disease patients from electronic health records. In: AMIA annual symposium proceedings, vol 2018. American Medical Informatics Association, p 340. https://www.ncbi.nlm.nih.gov/pmc/articles/PMC6371307/

Colbaugh R, Glass K, Rudolf C, Tremblay M (2018b) Robust ensemble learning to identify rare disease patients from electronic health records. In: 40th annual international conference of the IEEE engineering in medicine and biology society (EMBC). IEEE, pp 4085–4088. https://doi.org/10.1109/EMBC.2018.8513241

Colilla S, Tov EY, Zhang L, Kurzinger ML, Tcherny-Lessenot S, Penfornis C, Jen S, Gonzalez DS, Caubel P, Welsh S, Juhaeri J (2017) Validation of new signal detection methods for web query log data compared to signal detection algorithms used with FAERS. Drug Safety 40(5):399–408. https://doi.org/10.1007/s40264-017-0507-4

Cong Q, Feng Z, Li F, Zhang L, Rao G, Tao C (2018) Constructing biomedical knowledge graph based on SemMedDB and Linked Open Data. In: 2018 IEEE international conference on bioinformatics and biomedicine (BIBM), pp 1628–1631. https://doi.org/10.1109/BIBM.2018.8621568

Dandala B, Joopudi V, Devarakonda M (2019) Adverse drug events detection in clinical notes by jointly modeling entities and relations using neural networks. Drug Safety 42(1):135–146. https://doi.org/10.1007/s40264-018-0764-x

Deng Y, Li Y, Shen Y, Du N, Fan W, Yang M, Lei K (2019) MedTruth: A semi-supervised approach to discovering knowledge condition information from multi-source medical data. In: Proceedings of the 28th ACM international conference on information and knowledge management, pp 719–728. https://doi.org/10.1145/3357384.3357934

Devlin J, Chang MW, Lee K, Toutanova K (2018) Bert: Pre-training of deep bidirectional transformers for language understanding. In: Proceedings of NAACL-HLT 2019, pp 4171–4186. https://www.aclweb.org/anthology/N19-1423.pdf

Faruqui M, Dodge J, Jauhar SK, Dyer C, Hovy E, Smith NA (2015) Retrofitting word vectors to semantic lexicons. https://doi.org/10.3115/v1/N15-1184

Fauqueur J, Thillaisundara A, Togia T (2019) Constructing large scale biomedical knowledge bases from scratch with rapid annotation of interpretable patterns. https://arxiv.org/abs/1907.01417

FDA (2018) CDER conversation: Monitoring social media to better understand drug use trends. https://www.fda.gov/drugs/news-events-human-drugs/cder-conversation-monitoring-social-media-better-understand-drug-use-trends

Finlayson SG, LePendu P, Shah NH (2014) Building the graph of medicine from millions of clinical narratives. J Sci Data. https://doi.org/10.5061/dryad.jp917

Freifeld CC, Brownstein JS, Menone CM, Bao W, Filice R, Kass-Hout T, Dasgupta N (2014) Digital drug safety surveillance: Monitoring pharmaceutical products in Twitter. Drug Safety 37(5):343–350. https://doi.org/10.1007/s40264-014-0155-x

Garg R, Dong S, Shah S, Jonnalagadda SR (2016) A bootstrap machine learning approach to identify rare disease patients from electronic health records. https://arxiv.org/abs/1609.01586

Gavrielov-Yusim N, Kürzinger ML, Nishikawa C, Pan C, Pouget J, Epstein LBH, Golant Y, Tcherny-Lessenot S, Lin S, Hamelin B, Juhaeri J (2019) Comparison of text processing methods in social media-based signal detection. PDS Pharmacoepidemiol Drug Saf 28(10):1309–1317. https://doi.org/10.1002/pds.4857

Goodwin TR, Harabagiu SM (2016) Medical question answering for clinical decision support. In: Proceedings of the 25th ACM international conference on information and knowledge management, pp 297–306. https://www.ncbi.nlm.nih.gov/pmc/articles/PMC5530755/

Gunar G, Kukar M, Notar M, Brvar M, ernel P, Notar M, Notar M (2018) An application of machine learning to haematological diagnosis. Sci Rep 8(1). https://doi.org/10.1038/s41598-017-18564-8

Guo S, Wang Q, Wang B, Wang L, Guo L (2015) Semantically smooth knowledge graph embedding. In: Proceedings of the 53rd annual meeting of the association for computational linguistics and the 7th international joint conference on natural language processing (Volume 1: Long Papers), pp 84–94. https://www.aclweb.org/anthology/P15-1009.pdf

Hadzi-Puric J, Grmusa J (2012) Automatic drug adverse reaction discovery from parenting websites using disproportionality methods. In: 2012 IEEE/ACM international conference on advances in social networks analysis and mining, pp 792–797. https://doi.org/10.1109/ASONAM.2012.144

Han L, Ball R, Pamer CA, Altman RB, Proestel S (2017) Development of an automated assessment tool for MedWatch reports in the FDA adverse event reporting system. J Am Med Inform Assn 24(5):913–920. https://doi.org/10.1093/jamia/ocx022

Hinton GE, Salakhutdinov RR (2009) Replicated softmax: An undirected topic model. In: Advances in neural information processing systems, pp 1607–1614. http://papers.nips.cc/paper/3856-replicated-softmax-an-undirected-topic-model

Hofmann T (2001) Unsupervised learning by probabilistic latent semantic analysis. Mach Learn 42:177–196. https://doi.org/10.1023/A:1007617005950

Huang Z, Xu W, Yu K (2015) Bidirectional LSTM-CRF models for sequence tagging. https://arxiv.org/abs/1508.01991

Ibrahim M, Gauch S, Salman O, Alqahatani M (2020) Enriching consumer health vocabulary using enhanced GloVe word embedding. https://arxiv.org/ftp/arxiv/papers/2004/2004.00150.pdf

Jagannatha A, Liu F, Liu W, Yu H (2019) Overview of the first natural language processing challenge for extracting medication, indication, and adverse drug events from electronic health record notes (MADE 1.0). Drug Safety 42(1):99–111. https://doi.org/10.1007/s40264-018-0762-z

Ji S, Pan S, Cambria E, Marttinen P, Yu PS (2020) A survey on knowledge graphs: Representation, acquisition and applications. https://arxiv.org/abs/2002.00388

Jiang K, Zheng Y (2013) Mining Twitter data for potential drug effects. In: Motoda H, Wu Z, Cao L, Zaiane O, Yao M, Wang W (eds) Advanced data mining and applications. Springer, Heidelberg, pp 434–443. https://doi.org/10.1007/978-3-642-53914-5_37

Johnson MP, Johnson JC, Engel-Nitz NM, Said Q, Prestifilippo J, Gipson TT, Wheless J (2017) Management of a rare disease population: A model for identifying a patient population with tuberous sclerosis complex. Manag Care. https://pubmed.ncbi.nlm.nih.gov/28895825/

Joulin A, Grave E, Bojanowski P, Mikolov T (2017) Bag of tricks for efficient text classification. In: Lapata M, Blunsom P, Koller A (eds) Proceedings of the 15th conference of the European chapter of the association for computational linguistics: Volume 2, Short Papers. Association for Computational Linguistics, pp 427–431. https://doi.org/10.18653/v1/E17-2068

Kempf L, Goldsmith JC, Temple R (2018) Challenges of developing and conducting clinical trials in rare disorders. Am J Med Genet A 176(4):773–783. https://doi.org/10.1002/ajmg.a.38413

Kürzinger ML, Schück S, Texier N, Abdellaoui R, Faviez C, Pouget J, Zhang L, Tcherny-Lessenot S, Lin S, Juhaeri J (2018) Web-based signal detection using medical forums data in France: Comparative analysis. J Med Internet Res 20(11):e10,466. https://doi.org/10.2196/10466

Lavertu A, Altman RB (2019) RedMed: Extending drug lexicons for social media applications. J Biomed Inform 99:103,307. https://doi.org/10.1016/j.jbi.2019.103307

Leaman R, Wojtulewicz L, Sullivan R, Skariah A, Yang J, Gonzalez G (2010) Towards Internet-age pharmacovigilance: Extracting adverse drug reactions from user posts in health-related social networks. In: Proceedings of the 2010 workshop on biomedical natural language processing. Association for Computational Linguistics, pp 117–125. https://www.aclweb.org/anthology/W10-1915

Lee S, Han J, Park RW, Kim GJ, Rim JH, Cho J, Lee KH, Lee J, Kim S, Kim JH (2019) Development of a controlled vocabulary-based adverse drug reaction signal dictionary for multicenter electronic health record-based pharmacovigilance. Drug Safety 42(5):657–670. https://doi.org/10.1007/s40264-018-0767-7

Lengerich BJ, Maas AL, Potts C (2017) Retrofitting distributional embeddings to knowledge graphs with functional relations. https://arxiv.org/abs/1708.00112

LePendu P, Iyer SV, Bauer-Mehren A, Harpaz R, Mortensen JM, Podchiyska T, Ferris TA, Shah NH (2013) Pharmacovigilance using clinical notes. Clin Pharmac Ther 93(6):547–555. https://doi.org/10.1038/clpt.2013.47

Li F, Liu W, Yu H (2018) Extraction of information related to adverse drug events from electronic health record notes: Design of an end-to-end model based on deep learning. JMIR Med Inform 6(4):e12,159. https://doi.org/10.2196/12159

Li X, Wang Y, Wang D, Yuan W, Peng D, Mei Q (2019) Improving rare disease classification using imperfect knowledge graph. BMC Med Inform Decis 19(5):238. https://doi.org/10.1186/s12911-019-0938-1

Li L, Wang P, Yan J, Wang Y, Li S, Jiang J, Sun Z, Tang B, Chang TH, Wang S, Liu Y (2020) Real-world data medical knowledge graph: Construction and applications. Artif Intell Med 103:101,817. https://doi.org/10.1016/j.artmed.2020.101817

Lin Y, Liu Z, Luan H, Sun M, Rao S, Liu S (2015a) Modeling relation paths for representation learning of knowledge bases. https://arxiv.org/pdf/1506.00379.pdf

Lin Y, Liu Z, Sun M, Liu Y, Zhu X (2015b) Learning entity and relation embeddings for knowledge graph completion. In: Twenty-ninth AAAI conference on artificial intelligence

Liu H, Wu Y, Yang Y (2017) Analogical inference for multi-relational embeddings. In: Proceedings of the 34th international conference on machine learning, vol 70, pp 2168–2178. https://dl.acm.org/doi/10.5555/3305890.3305905

Liu Z, Peng E, Yan S, Li G, Hao T (2018) T-Know: A knowledge graph-based question answering and information retrieval system for traditional Chinese medicine. In: Proceedings of the 27th international conference on computational linguistics: system demonstrations, pp 15–19. https://www.aclweb.org/anthology/C18-2004.pdf

Liu F, Jagannatha A, Yu H (2019) Towards drug safety surveillance and pharmacovigilance: Current progress in detecting medication and adverse drug events from electronic health records. Drug Safety 42(1):95–97. https://doi.org/10.1007/s40264-018-0766-8

Magrabi F, Ammenwerth E, McNair JB, De Keizer NF, Hyppönen H, Nykänen P, Rigby M, Scott PJ, Vehko T, Wong ZSYJYomi (2019) Artificial intelligence in clinical decision support: Challenges for evaluating AI and practical implications. Yearb Med Inform 28(1):128–134. https://doi.org/10.1055/s-0039-1677903

Matsunaga D, Suzumura T, Takahashi T (2019) Exploring graph neural networks for stock market predictions with rolling window analysis. https://arxiv.org/pdf/1909.10660.pdf

Middleton B, Sittig DF, Wright A (2016) Clinical decision support: a 25 year retrospective and a 25 year vision. Yearb Med Inform 25(S01):S103–S116. https://doi.org/10.15265/IYS-2016-s034

Mikolov T, Sutskever I, Chen K, Corrado GS, Dean J (2013) Distributed representations of words and phrases and their compositionality. In: Proceedings of the 26th international conference on neural information processing systems, vol 2, pp 3111–3119. https://dl.acm.org/doi/10.5555/2999792.2999959

Mimno D, Wallach HM, Talley EM, Talley E, Leenders M, McCallum A (2011) Optimizing semantic coherence in topic models. In: Proceedings of the conference on empirical methods in natural language processing. Association for Computational Linguistics, pp 262–272. https://dl.acm.org/doi/10.5555/2145432.2145462

Mishra A, Laha A, Sankaranarayanan K, Jain P, Krishnan S (2019) Storytelling from structured data and knowledge graphs: An NLG perspective. In: Proceedings of the 57th annual meeting of the association for computational linguistics: Tutorial Abstracts, pp 43–48. https://www.aclweb.org/anthology/P19-4009.pdf

Newman D, Lau JH, Grieser K, Baldwin T (2010) Automatic evaluation of topic coherence. In: Human language technologies: The 2010 annual conference of the North American chapter of the Association for Computational Linguistics. Association for Computational Linguistics, pp 100–108. https://dl.acm.org/doi/10.5555/1857999.1858011

Ngo DH, Metke-Jimenez A, Nguyen A (2018) Knowledge-based feature engineering for detecting medication and adverse drug events from electronic health records. In: Proceedings of the 1st international workshop on medication and adverse drug event detection, vol 90, pp 31–38. http://proceedings.mlr.press/v90/ngo18a.html

Nguyen DQ, Billingsley R, Du L, Johnson M (2015) Improving topic models with latent feature word representations. T Assoc Comput Linguist 3:299–313. https://doi.org/10.1162/tacl_a_00140

Nickel M, Rosasco L, Poggio T (2016) Holographic embeddings of knowledge graphs. In: Proceedings of the thirtieth AAAI conference on artificial intelligence, pp 1955–1961. https://dl.acm.org/doi/10.5555/3016100.3016172

Nikfarjam A, Gonzalez GH (2011) Pattern mining for extraction of mentions of adverse drug reactions from user comments. In: AMIA annual symposium proceedings, vol 2011, pp 1019–1026. https://pubmed.ncbi.nlm.nih.gov/22195162

Nikfarjam A, Ransohoff JD, Callahan A, Jones E, Loew B, Kwong BY, Sarin KY, Shah NH (2019) Early detection of adverse drug reactions in social health networks: A natural language processing pipeline for signal detection. JMIR Public Health Surveill 5(2):e11,264. https://doi.org/10.2196/11264

Nordon G, Koren G, Shalev V, Kimelfeld B, Shalit U, Radinsky K (2019) Building causal graphs from medical literature and electronic medical records. In: Proceedings of the AAAI Conference on Artificial Intelligence, vol 33, pp 1102–1109. https://doi.org/10.1609/aaai.v33i01.33011102

O'Connor K, Pimpalkhute P, Nikfarjam A, Ginn R, Smith KL, Gonzalez G (2014) Pharmacovigilance on Twitter? Mining tweets for adverse drug reactions. AMIA Annu Symp Proc 2014:924–933. https://pubmed.ncbi.nlm.nih.gov/25954400

Pappa D, Stergioulas LK (2019) Harnessing social media data for pharmacovigilance: a review of current state of the art, challenges and future directions. Int J Data Sci Anal 8(2):113–135, https://doi.org/10.1007/s41060-019-00175-3

Pennington J, Socher R, Manning CD (2014) GloVe: Global vectors for word representation. In: Moschitti A, Pang B, Daelemans W (eds) Proceedings of the 2014 conference on empirical methods in natural language processing. Association for Computational Linguistics, pp 1532–1543. https://doi.org/10.3115/v1/D14-1162

Perera S, Henson C, Thirunarayan K, Sheth A, Nair S (2012) Data driven knowledge acquisition method for domain knowledge enrichment in the healthcare. In: 2012 IEEE international conference on bioinformatics and biomedicine. https://doi.org/10.1109/BIBM.2012.6392669

Perera S, Henson C, Thirunarayan K, Sheth A, Nair S (2013a) Semantics driven approach for knowledge acquisition from EMRs. IEEE J Biomed Health 18(2):515–524. https://doi.org/10.1109/JBHI.2013.2282125

Perera S, Sheth A, Thirunarayan K, Nair S, Shah N (2013b) Challenges in understanding clinical notes: Why NLP engines fall short and where background knowledge can help. In: Proceedings of the 2013 international workshop on data management & analytics for healthcare, pp 21–26. https://doi.org/10.1145/2512410.2512427

Peters ME, Neumann M, Iyyer M, Gardner M, Clark C, Lee K, Zettlemoyer L (2018) Deep contextualized word representations. https://arxiv.org/abs/1802.05365

Pierce CE, Bouri K, Pamer C, Proestel S, Rodriguez HW, Van Le H, Freifeld CC, Brownstein JS, Walderhaug M, Edwards IR, Dasgupta N (2017) Evaluation of Facebook and Twitter monitoring to detect safety signals for medical products: An analysis of recent FDA safety alerts. Drug Safety 40(4):317–331. https://doi.org/10.1007/s40264-016-0491-0

Pogue RE, Cavalcanti DP, Shanker S, Andrade RV, Aguiar LR, de Carvalho JL, Costa FF (2018) Rare genetic diseases: Update on diagnosis, treatment and online resources. Drug Discov Today 23(1):187–195, https://doi.org/10.1016/j.drudis.2017.11.002

Rehurek R, Sojka P (2010) Software framework for topic modelling with large corpora. In: Proceedings of the LREC 2010 workshop on new challenges for NLP frameworks, pp 46–50. https://doi.org/10.13140/2.1.2393.1847

Rotmensch M, Halpern Y, Tlimat A, Horng S, Sontag D (2017) Learning a health knowledge graph from electronic medical records. Sci Rep 7(1), https://doi.org/10.1038/s41598-017-05778-z

Ruan T, Huang Y, Liu X, Xia Y, Gao J (2019) QAnalysis: A question-answer driven analytic tool on knowledge graphs for leveraging electronic medical records for clinical research. BMC Med Inform Decis 19(1):82, https://doi.org/10.1186/s12911-019-0798-8

Sampathkumar H, Chen Xw, Luo B (2014) Mining adverse drug reactions from online healthcare forums using hidden Markov model. BMC Med Inform Decis 14(1):91, https://doi.org/10.1186/1472-6947-14-91

Sarker A, Ginn R, Nikfarjam A, O'Connor K, Smith K, Jayaraman S, Upadhaya T, Gonzalez G (2015) Utilizing social media data for pharmacovigilance: A review. J Biomed Inform 54:202–212. https://doi.org/10.1016/j.jbi.2015.02.004

Savova GK, Masanz JJ, Ogren PV, Zheng J, Sohn S, Kipper-Schuler KC, Chute CG (2010) Mayo clinical text analysis and knowledge extraction system (cTAKES): Architecture, component evaluation and applications. J Am Med Inform Assn 17(5):507–513. https://doi.org/10.1136/jamia.2009.001560

Schneider JM, Rehm G, Montiel-Ponsoda E, Rodriguez-Doncel V, Revenko A, Karampatakis S, Khvalchik M, Sageder C, Gracia J, Maganza F (2020) Orchestrating NLP services for the legal domain. In: Proceedings of The 12th language resources and evaluation conference, pp 2332–2340. https://www.aclweb.org/anthology/2020.lrec-1.284

Segura-Bedmar I MP Revert R (2014) Detecting drugs and adverse events from Spanish health social media streams. In: Proceedings of the 5th international workshop on health text mining and information analysis, pp 106–115. https://doi.org/10.3115/v1/W14-1117

Sethi T, Shah NH (2017) Pharmacovigilance using textual data: The need to go deeper and wider into the con(text). Drug Safety 40(11):1047–1048. https://doi.org/10.1007/s40264-017-0585-3

Socher R, Bauer J, Manning CD, Ng AY (2013a) Parsing with compositional vector grammars. In: Schuetze H, Fung P, Poesio M (eds) Proceedings of the 51st annual meeting of the association for computational linguistics (vol 1: Long papers). Association for Computational Linguistics, pp 455–465. https://www.aclweb.org/anthology/P13-1045.pdf

Socher R, Perelygin A, Wu J, Chuang J, Manning CD, Ng A, Potts C (2013b) Recursive deep models for semantic compositionality over a sentiment treebank. In: Yarowsky D, Baldwin T, Korhonen A, Livescu K, Bethard S (eds) Proceedings of the 2013 conference on empirical methods in natural language processing. Association for Computational Linguistics, pp 1631–1642. https://www.aclweb.org/anthology/D13-1170.pdf

Sohn S, Clark C, Halgrim SR, Murphy SP, Chute CG, Liu H (2014) MedXN: An open source medication extraction and normalization tool for clinical text. J Am Med Inform Assn 21(5):858–865. https://doi.org/10.1136/amiajnl-2013-002190

Sondhi P, Sun J, Tong H, Zhai C (2012) SympGraph: A framework for mining clinical notes through symptom relation graphs. In: Proceedings of the 18th ACM SIGKDD international conference on knowledge discovery and data mining, pp 1167–1175. https://doi.org/10.1145/2339530.2339712

Speer R, Chin J, Havasi C (2017) Conceptnet 5.5: An open multilingual graph of general knowledge. In: Thirty-first AAAI conference on artificial intelligence. https://dl.acm.org/doi/10.5555/3298023.3298212

Srivastava N, Salakhutdinov R, Hinton G (2013) Fast inference and learning for modeling documents with a deep Boltzmann machine. In: International conference on machine learning (ICML) workshop on inferning: Interactions between inference and learning. https://openreview.net/pdf?id=GtacG-v9TXtUf

Stergiopoulos S, Fehrle M, Caubel P, Tan L, Jebson L (2019) Adverse drug reaction case safety practices in large biopharmaceutical organizations from 2007 to 2017: An industry survey. Pharm Med 33(6):499–510. https://doi.org/10.1007/s40290-019-00307-x

Suárez-Figueroa MC, Gómez-Pérez A, Fernández-López M (2012) The NeOn methodology for ontology engineering. In: Suárez-Figueroa MC, Gómez-Pérez A, Motta E, Gangemi A (eds) Ontology engineering in a networked world. Springer, Heidelberg, pp 9–34. https://doi.org/10.1007/978-3-642-24794-1_2

Tao Y, Godefroy B, Genthial G, Potts C (2018) Effective feature representation for clinical text concept extraction. https://arxiv.org/pdf/1811.00070.pdf

Teh YW, Jordan MI, Beal MJ, Blei DM (2004) Sharing clusters among related groups: Hierarchical Dirichlet processes. In: Advances in neural information processing systems, pp 1385–1392. http://papers.nips.cc/paper/2698-sharing-clusters-among-related-groups-hierarchical-dirichlet-processes.pdf

Throckmorton DC, Gottlieb S, Woodcock J (2018) The FDA and the next wave of drug abuse — proactive pharmacovigilance. N Engl J Med 379(3):205–207, https://doi.org/10.1056/NEJMp1806486

Tong P, Zhang Q, Yao JJDS, Engineering (2019) Leveraging domain context for question answering over knowledge graph. Data Sci Eng 4(4):323–335, https://doi.org/10.1007/s41019-019-00109-w

Torii M, Wagholikar K, Liu H (2011) Using machine learning for concept extraction on clinical documents from multiple data sources. J Am Med Inform Assn 18(5):580–587. https://doi.org/10.1136/amiajnl-2011-000155

van Stekelenborg J, Ellenius J, Maskell S, Bergvall T, Caster O, Dasgupta N, Dietrich J, Gama S, Lewis D, Newbould V, Brosch S, Pierce CE, Powell G, Ptaszyska-Neophytou A, Winiewski AFZ, Tregunno P, Norén GN, Pirmohamed M (2019) Recommendations for the use of social media in pharmacovigilance: Lessons from IMI WEB-RADR. Drug Safety 42(12):1393–1407, https://doi.org/10.1007/s40264-019-00858-7

Wang Y, Cai Y (2017) Message passing on factor graph: A novel approach for orphan drug physician targeting. In: Perner P (ed) Advances in data mining. applications and theoretical aspects. Springer, Cham, pp 137–150. https://doi.org/10.1007/978-3-319-62701-4_11

Wang Z, Zhang J, Feng J, Chen Z (2014) Knowledge graph embedding by translating on hyperplanes. In: Twenty-eighth AAAI conference on artificial intelligence, pp 1112–1119. https://dl.acm.org/doi/abs/10.5555/2893873.2894046

Wang L, Rastegar-Mojarad M, Ji Z, Liu S, Liu K, Moon S, Shen F, Wang Y, Yao L, Davis III JM, Liu H (2018) Detecting pharmacovigilance signals combining electronic medical records with spontaneous reports: A case study of conventional disease-modifying antirheumatic drugs for rheumatoid arthritis. Front Phramacol 9(875). https://doi.org/10.3389/fphar.2018.00875

Wasylewicz ATM, Scheepers-Hoeks AMJW (2019) Clinical decision support systems. In: Kubben P, Dumontier M, Dekker A (eds) Fundamentals of clinical data science. Springer, Cham, pp 153–169. https://doi.org/10.1007/978-3-319-99713-1_11

Xia E, Sun W, Mei J, Xu E, Wang K, Qin Y (2018) Mining disease-symptom relation from massive biomedical literature and its application in severe disease diagnosis. In: AMIA annual symposium proceedings, Am. Med. Inform. Assn., vol 2018, p 1118. https://www.ncbi.nlm.nih.gov/pmc/articles/PMC6371303/

Xu K, Reddy S, Feng Y, Huang S, Zhao D (2016) Question answering on Freebase via relation extraction and textual evidence. https://arxiv.org/pdf/1603.00957.pdf

Xu B, Xu Y, Liang J, Xie C, Liang B, Cui W, Xiao Y (2017) CN-DBpedia: A never-ending Chinese knowledge extraction system. In: International conference on industrial, engineering and other applications of applied intelligent systems. Springer, Cham, pp 428–438. https://doi.org/10.1007/978-3-319-60045-1_44

Yang CC, Yang H, Jiang L, Zhang M (2012) Social media mining for drug safety signal detection. In: Proceedings of the 2012 international workshop on smart health and wellbeing, pp 33–40. https://doi.org/10.1145/2389707.2389714

Yang B, Yih Wt, He X, Gao J, Deng L (2014a) Embedding entities and relations for learning and inference in knowledge bases. https://arxiv.org/pdf/1412.6575.pdf

Yang CC, Yang H, Jiang L (2014b) Postmarketing drug safety surveillance using publicly available health-consumer-contributed content in social media. ACM Trans Manage Inf Syst 5(1):Article 2. https://doi.org/10.1145/2576233

Yang X, Bian J, Gong Y, Hogan WR, Wu Y (2019) MADEx: A system for detecting medications, adverse drug events, and their relations from clinical notes. Drug Safety 42(1):123–133. https://doi.org/10.1007/s40264-018-0761-0

Yao L, Zhang Y, Wei B, Jin Z, Zhang R, Zhang Y, Chen Q (2017) Incorporating knowledge graph embeddings into topic modeling. In: Thirty-first AAAI conference on artificial intelligence. https://dl.acm.org/doi/abs/10.5555/3298483.3298687

Yates A, Goharian N (2013) ADRTrace: Detecting expected and unexpected adverse drug reactions from user reviews on social media sites. In: Serdyukov P, Braslavski P, Kuznetsov SO, Kamps J, Rüger S, Agichtein E, Segalovich I, Yilmaz E (eds) Advances in information retrieval. Springer, Heidelberg, pp 816–819. https://doi.org/10.1007/978-3-642-36973-5_92

Yeleswarapu S, Rao A, Joseph T, Saipradeep VG, Srinivasan R (2014) A pipeline to extract drug-adverse event pairs from multiple data sources. BMC Med Inform Decis 14(1):13. https://doi.org/10.1186/1472-6947-14-13

Yu Y, Ruddy KJ, Hong N, Tsuji S, Wen A, Shah ND, Jiang G (2019) ADEpedia-on-OHDSI: A next-generation pharmacovigilance signal detection platform using the OHDSI common data model. J Biomed Inform 91:103,119. https://doi.org/10.1016/j.jbi.2019.103119

Yuan J, Jin Z, Guo H, Jin H, Zhang X, Smith T, Luo J (2020) Constructing biomedical domain-specific knowledge graph with minimum supervision. Knowl Inf Syst 62(1):317–336. https://doi.org/10.1007/s10115-019-01351-4

Zhang Y, Dai H, Kozareva Z, Smola AJ, Song L (2017) Variational reasoning for question answering with knowledge graph. https://arxiv.org/abs/1709.04071

Zheng W, Zhang M (2019) Question answering over knowledge graphs via structural query patterns. https://arxiv.org/pdf/1910.09760.pdf

Zheng W, Yu JX, Zou L, Cheng H (2018) Question answering over knowledge graphs: question understanding via template decomposition. Proc VLDB Endowment 11(11):1373–1386. https://doi.org/10.14778/3236187.3236192

Zhou L, Plasek JM, Mahoney LM, Karipineni N, Chang F, Yan X, Chang F, Dimaggio D, Goldman DS, Rocha RA (2011) Using medical text extraction, reasoning and mapping system (MTERMS) to process medication information in outpatient clinical notes. In: AMIA annual symposium proceedings, vol 2011, pp 1639–1648. https://pubmed.ncbi.nlm.nih.gov/22195230. https://www.ncbi.nlm.nih.gov/pmc/articles/PMC3243163/

Printed in the United States
by Baker & Taylor Publisher Services